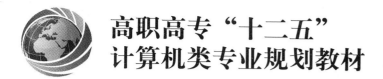

高职高专"十二五"
计算机类专业规划教材

Android 应用开发基础教程

主　编　贺　维　朱　俭
副主编　刘丽涛　潘　艺　宋春晖
编　写　李　鑫　王丽红　邱文严　孙兴华
主　审　吴程林

内 容 提 要

本书共 8 章，详细讲解了移动通信技术简介、Android 程序设计基础、Android 用户界面开发、Android 手机功能开发、Android 数据存储开发、Android 文件管理、Android 多媒体开发、Android 游戏开发等基础知识，读者可以根据自身的需要进行学习。本书在讲解过程中，对一些基础知识给出了实际的程序代码，可以使读者很快掌握知识点的应用。

全书采用"任务驱动式"教学方法，使学生带着问题学习，学习目标更加明确，通过本书的学习，学生能够在较短的时间内掌握 Android 开发技术。本书内容丰富，结构清晰，图文并茂，语言简练，通俗易懂，充分考虑到初学者的需要，具有较强的实用性和可操作性。

本书可作为高职高专院校 Android 开发技术课程教材，也适合具备 Java 基础及一定软件开发基础知识、具备一些手机开发经验的开发者和 Android 开发爱好者学习使用，还可作为相关培训学校的 Android 培训教材。

图书在版编目（CIP）数据

Android 应用开发基础教程 / 贺维，朱俭主编. —北京：中国电力出版社，2014.11
高职高专"十二五"计算机类专业规划教材
ISBN 978-7-5123-6634-3

Ⅰ.①A… Ⅱ.①贺… ②朱… Ⅲ.①移动终端－应用程序－程序设计－高等职业教育－教材 Ⅳ.①TN929.53

中国版本图书馆 CIP 数据核字（2014）第 238841 号

中国电力出版社出版、发行
（北京市东城区北京站西街 19 号　100005　http://www.cepp.sgcc.com.cn）
北京丰源印刷厂印刷
各地新华书店经售

＊

2014 年 11 月第一版　　2014 年 11 月北京第一次印刷
787 毫米×1092 毫米　16 开本　10.5 印张　249 千字
定价 **21.00** 元

敬 告 读 者

本书封底贴有防伪标签，刮开涂层可查询真伪
本书如有印装质量问题，我社发行部负责退换
版 权 专 有　翻 印 必 究

前 言

Android 是一个优秀的开源手机平台。本书以 Android 2.3.3 为系统平台，使用 Eclipse 为开发工具，介绍在 Android 平台上进行应用开发的知识和技术。本书从教学的角度全面介绍了 Android 应用程序的开发设计，深浅适宜，实例丰富，不仅可作为高职高专和计算机培训机构的相关课程的教材，而且也可作为 Android 系统开发设计人员的参考书。

本书采用"任务驱动式"教学法，共分为 8 章，将移动通信技术简介、Android 程序设计基础、Android 用户界面开发、Android 手机功能开发、Android 数据存储开发、Android 文件管理、Android 多媒体开发、Android 游戏开发等技能点贯穿于工作任务之中。每个工作任务由需求分析、知识准备、任务实施三部分组成。每个任务介绍一个完整的知识点，具有一定的代表性和实践意义。

（1）需求分析：介绍工作任务，对工作任务的要求进行说明。
（2）知识准备：介绍工作任务中涉及的主要知识和技能点。
（3）任务实施：对完成任务需要的工作流程进行描述。

本书凝聚了作者多年的教学与手机开发经验，讲解深入透彻，论述通俗易懂，注重知识的系统性，案例解析清晰透彻。凡具备编程基础的人员都可以通过本书的学习掌握 Android 的应用编程。教学资源齐全包括电子课件、程序源代码、案例库及相关网络资源。

本书由贺维、朱俭任主编，刘丽涛、潘艺、宋春晖任副主编。各章编写分工如下：第 1、7 章由贺维编写，第 3 章由朱俭编写，第 2、6 章由刘丽涛编写，第 4、5 章由潘艺编写，第 8 章由宋春晖编写。参与编写的还有李鑫、王丽红、邱文严、孙兴华。

限于编者水平，加之创作时间仓促，本书难免有不足之处，欢迎广大读者批评指正。

编　者
2014 年 6 月

目 录

前言
第 1 章 移动通信技术简介 ……………………………………………………………… 1
引言 …………………………………………………………………………………………… 1
1.1 移动通信技术概述 ………………………………………………………………… 1
1.2 Android 技术简介 ………………………………………………………………… 5
第 2 章 Android 程序设计基础 ………………………………………………………… 10
引言 ………………………………………………………………………………………… 10
2.1 Android 开发环境搭建 …………………………………………………………… 10
2.2 Android 应用程序目录结构 ……………………………………………………… 19
2.3 Android 生命周期 ………………………………………………………………… 24
第 3 章 Android 用户界面开发 ………………………………………………………… 30
引言 ………………………………………………………………………………………… 30
3.1 需求分析 …………………………………………………………………………… 30
3.2 知识准备 …………………………………………………………………………… 30
3.3 任务实施 …………………………………………………………………………… 56
第 4 章 Android 手机功能开发 ………………………………………………………… 68
引言 ………………………………………………………………………………………… 68
4.1 Intent 组件的使用 ………………………………………………………………… 68
4.2 Android 广播机制 ………………………………………………………………… 74
4.3 Android 资源管理器 ……………………………………………………………… 76
4.4 手机基本功能开发 ………………………………………………………………… 84
第 5 章 Android 数据存储开发 ………………………………………………………… 91
引言 ………………………………………………………………………………………… 91
5.1 需求分析 …………………………………………………………………………… 91
5.2 知识准备 …………………………………………………………………………… 91
5.3 任务实施 …………………………………………………………………………… 98

第 6 章　Android 文件管理 ·· 104
　　引言 ··· 104
　　6.1　需求分析 ··· 104
　　6.2　知识准备 ··· 104
　　6.3　任务实施 ··· 107
第 7 章　Android 多媒体开发 ·· 116
　　引言 ··· 116
　　7.1　需求分析 ··· 116
　　7.2　知识准备 ··· 116
　　7.3　任务实施 ··· 126
第 8 章　Android 游戏开发 ·· 135
　　引言 ··· 135
　　8.1　需求分析 ··· 135
　　8.2　知识准备 ··· 135
　　8.3　任务实施（一）··· 139
　　8.4　任务实施（二）··· 147

参考文献 ··· 159

第 1 章 移动通信技术简介

引言

随着移动通信技术的不断发展，移动终端应用技术越来越受到用户的关注，用户的主要工作平台已从原有的 PC 级平台，逐步向智能终端平台转移，Android 平台应运而生，Android 平台以其优秀平台架构，友好的用户界面，高度的开发性，逐步被用户接受，目前已经成为市场占有率最高、应用最广的移动智能终端系统。本章主要介绍移动通信技术的基本发展历史，目前主流的 3G 应用技术，不同的智能终端平台的特点，通过不同智能平台的比较分析 Android 平台的技术优势，从理论上整体介绍 Android 平台的系统架构，构成 Android 应用程序的四大组件。

1.1 移动通信技术概述

1.1.1 移动通信技术的发展

1. 第一代移动通信技术

1995 年问世的第一代模拟制式手机（1G）只能进行语音通话，最大的特点体现在其移动性上，第一代移动通信系统是基于模拟传输的，其特点是业务量小、质量差、安全性差、没有加密和速度低。

2. 第二代移动通信技术

1996—1997 年出现的第二代 GSM、CDMA 等数字制式手机（2G）便增加了接收数据的功能，如接收电子邮件或网页，数字信号技术是其最基本的特征（包括 GSM 数字移动通信技术及窄带 CDMA），使用 SIM 卡、使用轻型手机和对大量用户的网络支撑能力。但随着用户规模和网络规模的不断扩大，频率资源已接近枯竭，语音质量不能达到用户满意的标准，数据通信速率太低，无法在真正意义上满足移动多媒体业务的需求。

3. 第三代移动通信技术

简称 3G，全称为 3rd Generation，第三代与前两代的主要区别是在传输声音和数据的速度上的提升，它能够在全球范围内更好地实现无缝漫游，并处理图像、音乐、视频流等多种媒体形式，提供包括网页浏览、电话会议、电子商务等多种信息服务，同时也要考虑与已有第二代系统的良好兼容性。

第三代移动通信的基本特征如下：

（1）具有全球范围设计的，与固定网络业务及用户互连，无线接口的类型尽可能少和高度兼容；

（2）具有与固定通信网络相比拟的高话音质量和高安全性；

（3）具有在本地采用 2Mb/s 高数据速率接入和在广域网采用 384kb/s 接入速率的数据速率分段使用功能；

（4）具有 2GHz 左右的高效频谱利用率，且能最大程度地利用有限带宽；

（5）移动终端可连接地面网和卫星网，可移动使用和固定使用，可与卫星业务共存和互连；

（6）能够处理包括国际互联网和视频会议、高数据率通信和非对称数据传输的分组和电路交换业务；

（7）支持分层小区结构，也支持包括用户向不同地点通信时浏览因特网的多种同步连接；

（8）语音只占移动通信业务的一部分，大部分业务是非话数据和视频信息；

（9）一个共用的基础设施，可支持同一地方的多个公共的和专用的运营公司；

（10）手机体积小、重量轻，具有真正的全球漫游能力；

（11）具有根据数据量、服务质量和使用时间为收费参数，而不是以距离为收费参数的新收费机制。

4. 第四代移动通信技术

4G（第四代移动通信技术）的概念可称为宽带接入和分布网络，具有非对称的超过 2Mb/s 的数据传输能力。它包括宽带无线固定接入、宽带无线局域网、移动宽带系统和交互式广播网络。第四代移动通信标准比第三代标准具有更多的功能。第四代移动通信可以在不同的固定、无线平台和跨越不同的频带的网络中提供无线服务，可以在任何地方用宽带接入互联网（包括卫星通信和平流层通信），能够提供定位定时、数据采集、远程控制等综合功能。此外，第四代移动通信系统是集成多功能的宽带移动通信系统，是宽带接入 IP 系统。

第四代移动通信技术的主要指标：

（1）数据速率从 2Mb/s 提高到 100Mb/s，移动速率从步行到车速以上。

（2）支持高速数据和高分辨率多媒体服务的需要。宽带局域网应能与 B-ISDN 和 ATM 兼容，实现宽带多媒体通信，形成综合宽带通信网。

（3）对全速移动用户能够提供 150Mb/s 的高质量影像等多媒体业务。

1.1.2 主流 3G 技术标准

表 1-1 所示为目前主流的 3G 技术标准，而国内三大移动运营商，分别对不同的 3G 技术标准进行支持。中国联通对于 WCDMA 标准进行支持，目前 WCDMA 的技术比较成熟，网络环境较好，在 3G 市场竞争中占据一定优势。但随着电信凭借它的雄厚实力，对 CDMA 2000 技术标准的大规模推进和低价策略，中国电信的 3G 市场份额正在逐渐增大，目前，中国电信是全球最大的 CDMA 网络运营商。中国移动所持有的 TD-SCDMA 技术为国家标准，但是在技术上存在一定差距，技术服务和网络环境相对较差，而中国移动手中有大量 GSM 用户的丰厚利润，因此中国移动对于 3G 业务推广激情不高，中国移动的工作重点目前放在了 4G 技术方面。目前国标 4G 技术已经接近商用，有十几个国家已经宣布支持中国的 4G 标准。

表 1-1　　　　　　　　　　　　3G 技 术 标 准

技术	TD-SCDMA （中国移动）	CDMA 2000 （中国电信）	WCDMA （中国联通）
系统成熟度	较弱	一般	较强
全球漫游能力	较弱	一般	较强
优化经验	较少	较强	较强
数据速率	下行：2.8Mb/s 上行：384kb/s	下行：3.1Mb/s 上行：1.8Mb/s	下行：14.4Mb/s 上行：5.76Mb/s
国内开展业务时间	较早	一般	较晚
高速移动性能	较弱	较强	较强
产业链成熟度	较弱	一般	较强
2G 现有用户群	很多	一般	较多
2G 3G 网络互操作	一般	较强	较强

1.1.3　主流移动终端平台

3G 技术看起来很美，但是，国内绝大部分用户仍然停留在 2G 应用中，使用 GSM 制式的手机。3G 技术的实质，就是在为用户提供更高速和安全的网络环境下，为用户提供更多样化的服务，但大多数用户除了用户升级成本因素外并没有感觉到 3G 技术的优势，其中，最主要的一个因素是 3G 技术的应用相对比较滞后，用户感受不到 3G 网络对我们生活的巨大影响。随着 3G 技术的不断推进，乃至 4G 技术的发展，技术的应用都是一个亟需解决的问题，让用户真正感受到技术进步对我们生活的改变。

3G 移动终端，帮助我们进行 3G 网络连接，移动终端可以是各种电器设备，目前最常见的移动终端设备就是手机，手机领域的应用也是目前各大公司关注的焦点，几乎所有软件和硬件公司都认识到，随着泛计算机时代的到来，随着技术的进步，手机技术和平板电脑技术将在很大程度上代替传统意义上的计算机，将是未来计算机技术发展的一个至关重要的领域。3G 移动终端技术将得到飞速的发展。

目前，各大公司都不会放弃这块巨大的蛋糕，纷纷推出了自己的 3G 移动终端平台，希望在未来的竞争中占据一个有利的位置，下面对目前主流的 3G 移动终端平台进行介绍。

1. Android

Android 是基于 Linux 内核的操作系统，是 Google 公司在 2007 年 11 月 5 日公布的手机操作系统，早期由原名为"Android"的公司开发，Google 在 2005 年对其进行收购，并命名为"Android.Inc"，Google 继续对 Android 系统开发运营，它采用了软件堆层（software stack，又名软件叠层）的架构。底层 Linux 内核只提供基本功能；其他的应用软件则由各公司自行开发，部分程序以 Java 编写。

Android 平台五大优势特色：

（1）开放性：在优势方面，Android 平台首先就是其开放性，开发的平台允许任何移动终端厂商加入到 Android 联盟中来。显著的开放性可以使其拥有更多的开发者，随着用户和应用的日益丰富，一个崭新的平台也将很快走向成熟。对于 Android 的发展而言，开放性有利于积累人气，这里的人气包括消费者和厂商，而对于消费者来讲，最大的受益正是丰富的软件资源。开放的平台也会带来更大竞争，如此一来，消费者将可以用更低的价位购得心仪的手机。

（2）挣脱运营商的束缚：在过去很长的一段时间，特别是在欧美地区，手机应用往往受到运营商制约，使用什么功能接入什么网络，几乎都受到运营商的控制。自从 iPhone 上市，用户可以更加方便地连接网络，运营商的制约减少。随着 EDGE、HSDPA 这些 2G 至 3G 移动网络的逐步过渡和提升，手机随意接入网络已不是运营商口中的笑谈。

（3）丰富的硬件选择：这一点还是与 Android 平台的开放性相关，由于 Android 的开放性，众多的厂商会推出千奇百怪，功能特色各具的多种产品。功能上的差异和特色，却不会影响到数据同步、甚至软件的兼容。好比你从诺基亚 Symbian 风格手机一下改用苹果 iPhone，同时还可将 Symbian 中优秀的软件带到 iPhone 上使用、联系人等资料更是可以方便地转移。

（4）不受任何限制的开发商：Android 平台提供给第三方开发商一个十分宽泛、自由的环境，因此不会受到各种条条框框的阻挠，可想而知，会有多少新颖别致的软件诞生。但也有其两面性，血腥、暴力、情色方面的程序和游戏如何控制正是留给 Android 的难题之一。

（5）无缝结合的 Google 应用：如今叱咤互联网的 Google 已经走过 10 年的历史。从搜索巨人到全面的互联网渗透，Google 服务如地图、邮件、搜索等已经成为连接用户和互联网的重要纽带，而 Android 平台手机将无缝结合这些优秀的 Google 服务。

2．IOS

IOS 是苹果公司为 iPhone 开发的操作系统，它主要是给 iPhone、iPod touch 及 iPad 使用。就像 IOS 基于的 Mac OS X 操作系统一样，它也是以 Darwin 为基础的。原本这个系统名为 iPhone OS，在 2010 年 6 月 7 日的 WWDC 大会上被改名为 iOS。iOS 的系统架构分为四个层次：核心操作系统层（the Core OS layer）、核心服务层（the Core Services layer）、媒体层（the Media layer）、可轻触层（the Cocoa Touch layer）。系统操作占用大概 240MB 的存储器空间。

3．Symbian

Symbian 操作系统对移动终端产品进行了最优化设计。Symbian 操作系统在智能移动终端上拥有强大的应用程序及通信能力，都要归功于它有一个非常健全的核心——强大的对象导向系统、企业用标准通信传输协议及完美的 Sunjava 语言。Symbian 认为无线通信装置除了要提供语音沟通的功能外，同时也应具有其他多种沟通方式，如触笔、键盘等。在硬件设计上，它可以提供许多不同风格的外形，像使用真实或虚拟的键盘；在软件功能上，它可以容纳许多功能，包括和他人互相分享信息、浏览网页、收发电子邮件、传真及个人生活行程管理等。此外，Symbian 操作系统在扩展性方面为制造商预留了多种接口。

4．Windows Phone

Windows Phone（简称 WP）是微软发布的一款手机操作系统，它将微软旗下的 Xbox Live 游戏、Xbox Music 音乐与独特的视频体验集成至手机中。Windows Phone 具有桌面定制、图标拖拽、滑动控制等一系列前卫的操作体验。其主屏幕通过提供类似仪表盘的体验来显示新的电子邮件、短信、未接来电、日历约会等，让人们对重要信息保持时刻更新。很容易看出微软在用户操作体验上所做出的努力——全新的 Windows 手机把网络、个人电脑和手机的优势集于一身，让人们可以随时随地享受到想要的体验。

5．Linux

Linux 手机操作系统是由计算机 Linux 操作系统变化而来的。简单地说，Linux 是一套免费使用和自由传播的操作系统,它具有稳定、可靠、安全等优点,有强大的网络功能。在相关软件的支持下,可实现 WWW、FTP、DNS、DHCP、E-mail 等服务。Linux 具有源代码开放这一特点非常重要,因为丰富的应用是智能手机的优越性体现和关键卖点所在。从应用开发的角度看,由于 Linux 的源代码是开放的,有利于独立软件开发商(ISV)开发出硬件利用效率高、功能更强大的应用软件,也方便行业用户开发自己的安全、可控认证系统。

6. Palm

PalmOS 属于 Palm 公司,是一种 32 位的嵌入式操作系统,它的操作界面采用触控式,几乎所有的控制选项都排列在屏幕上,使用触控笔便可进行所有操作。作为一套极具开放性的系统,开发商向用户免费提供 Palm 操作系统的开发工具,允许用户利用该工具在 Palm 操作系统的基础上编写、修改相关软件,使支持 Palm 的应用程序丰富多彩、应有尽有。Palm 操作系统最明显的优势还在于其本身是一套专门为掌上电脑编写的操作系统,在编写时充分考虑到了掌上电脑内存相对较小的情况,所以 Palm 操作系统本身所占的内存极小,基于 Palm 操作系统编写的应用程序所占的空间也很小,通常只有几十千字节,所以基于 Palm 操作系统的掌上电脑虽然只有几兆字节内存却可以运行众多的应用程序。Palm 在其他方面还存在一些不足,Palm 操作系统本身不具有录音、MP3 播放功能等,如果用户需要使用这些功能,就需要另外加入第三方软件或硬件设备方可实现。

1.2 Android 技术简介

1.2.1 Android 简史

2005 年 Google 收购了刚刚成立 22 个月的 Android 公司,开始从事智能操作系统的研发工作,图 1-1 为 Android 系统的标志。

为了能够对抗苹果公司在智能手机上的霸主地位,2007 年 11 月 5 日,以 Google 为首的 34 家公司宣布开发手机联盟(Open handset alliance,简称 OHA),该联盟受到广泛支持,联盟规模也在不断扩大,表 1-2 为 OHA 的主要成员。

图 1-1 Android 的标志

表 1-2 OHA 的主要成员

	电信运营商	软件开发商	商业公司	半导体公司	手机制造商
创始成员	中国移动 KDDI NTT DoCoMo Sprint Nextel T-Mobile 意大利电讯 Telefónica	Ascender eBay Esmertec Google LivingImage Myriad NMS Communications Nuance Communications PacketVideo SkyPop SONiVOX	Aplix Noser Engineering The Astonishing Tribe 风河系统	Audience 博通 英特尔 迈威尔科技 Nvidia 高通公司 SiRF 科技集团 Synaptics 德州仪器	HTC LG 摩托罗拉 三星电子

续表

	电信运营商	软件开发商	商业公司	半导体公司	手机制造商
2008年12月9日加入	Vodafone Softbank		Borqs Omron Software Teleca	AKM Semiconductor ARM Atheros Communications EMP	华硕电脑 Garmin 华为 索尼爱立信 爱立信 东芝
2009年5月17日加入	中国联通				
2009年5月27日加入		SVOX			
2009年6月1日加入					宏碁
2009年9月30日加入				MIPS科技公司	
2010年1月6日加入	中国电信				
2010年1月15日加入					中兴通信
2010年1月22日加入			Sasken Communication Technologies		
2010年7月12日加入				联发科技	

图1-2　T-mobile G1

2008年9月22日,T-mobile在纽约正式发布第一款Google手机——T-mobile G1,如图1-2所示。成为真正意义上的第一部基于Android操作系统的智能手机。

随着Android操作系统的不断完善,基于Android平台的终端设备层出不穷,目前Android软件获得飞速发展,Android手机的市场占有率逐年递增,已经成为当之无愧的3G移动终端智能平台的王者。Google为Android赋予的使命与Google本身的纲领是一致的,即随时随地为每一个人提供信息,如图1-3所示,为目前Android market中最流行的应用软件,可见绝大部分软件的目的都是在为用户提供及时有效的信息服务。

图1-3　Android market中最流行的应用软件

注：2013年9月4日凌晨，Google对外公布了Android新版本Android 4.4KitKat（奇巧巧克力），2013年09月24日Google开发的操作系统Android迎来了5岁生日，全世界采用这款系统的设备数量已经达到10亿台。Android平台手机的全球市场份额已经达到78.1%。

1.2.2　Android平台优势

IOS和Android作为最优秀的智能平台，都具有它不可比拟的优势。为了能够占有更大的市场份额，Google与苹果公司在移动智能平台上展开激烈的竞争，如图1-4所示。

图1-4　IOS VS Android

表1-3所示是Android与IOS系统主要功能的比较。苹果公司推出的IOS系统，对于软件开发公司来说，苹果平台下有非常清晰的盈利模式，因此，软件开发公司乐于进行IOS平台下的软件设计，因此IOS平台下拥有大量的应用软件资源，苹果对于IOS平台下的软件进行严格审核，保证了软件的基本质量。

表1-3　　　　　　　　　　Android与IOS系统功能比较

Android	iPhone
可以在PC，MAC和Linux下开发	只能在MAC下开发
以Linux为基础	以MAC OS为基础
Java	Objective C
25美元	每年99美元
可以通过Web下载应用程序	只能在App Store下载应用程序
支持Flash	不支持Flash
Google、ARM、高通、三星…	Apple
超过50 000个应用程序	超过100 000个应用程序

Android系统的优势主要表现在以下几个方面：

（1）Android价格占优：消费者选择产品，价格是必然要考虑的一大因素，iphone虽好，但是价格让一般人望而却步。

（2）应用程序发展迅速：智能机玩的就是应用，Android应用商店的推出，开发成本的低廉受到广大开发人员的支持。

（3）智能手机厂家助力、机型多：厂商加盟越多，手机终端就会越多，其市场潜力就越大，自从Google推出Android系统以来，各大厂家纷纷推出自己的Android平台手机，HTC、索尼爱立信、魅族、摩托罗拉、夏普、LG、三星、联想等，每一家手机厂商都推出了各自的

Android 手机。

（4）系统开源，利于创新：Android 是开源的，允许第三方修改，这在很大程度上允许厂家根据自己的硬件更改版本，从而能够更好的适应硬件，与之形成良好的结合。

1.2.3 Android 平台的结构

Android 平台，基于 Linux 内核，针对移动智能终端的特性，进行重新设计，图 1-5 所示为 Android 平台的结构图。

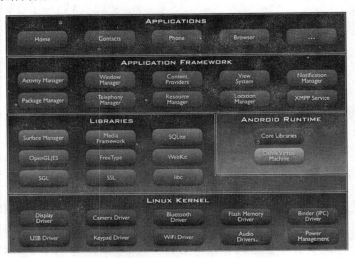

图 1-5 Android 平台的结构图

我们可以看出，Android 平台由 Linux Kernel、Libraries 和 Android Runtime、Application Framework、Applications 四部分组成。

（1）Linux Kernel：主要作用是实现与硬件的交互。Linux Kernel 基于 Linux 核心，并包含与硬件交互所需驱动程序，为平台提供最核心最基础的功能。

（2）Libraries 和 Android Runtime：提供 Android 应用程序运行时所依赖的各种程序库，并配置应用程序运行环境。Libraries 是使用 C/C++开发的程序包，为平台应用软件开发提供支持，Android 平台下的应用软件采用 Java 语言进行开发，我们知道要运行 Java 应用程序需要配置运行环境，即需要安装 Java 虚拟机，为了适应智能平台的需要，Google 重新优化设计了 Java 虚拟机，与 Google 特有的一些类库一起集成在 Android Runtime 中。

（3）Application Framework：提供 Android 应用程序开发所必需的基本框架，该层包含了移动终端所必需的 API，采用 Java 语言进行编写。

（4）Applications：用户开发的应用程序。

Android 平台应用软件运行的工作原理：

（1）应用程序运行时，会调用应用程序框架。

（2）应用程序框架根据应用程序不同的请求，分配虚拟机和访问指定程序库。

（3）虚拟机根据应用程序框架的请求，发出指令给 Linux Kernel，完成与硬件设备之间的交互。

1.2.4 Android 应用程序的四大组件

应用程序的核心是进行数据的操作，Android 应用程序为了能够更好的实现移动终端平台下的数据操作，设计了 4 个最重要的应用程序组建，即：activity、intent、service、content provider。

（1）activity：就是应用程序的界面，负责应用程序数据的展示，一般每个界面就是一个 activity；

（2）intent：传输作用，android 应用程序中所有数据都通过这个组件进行传输；

（3）service：提供服务支持，负责数据处理的工作，程序不可见；

（4）content provider：存储数据，并允许有需要的应用程序访问这些数据，相当于 Android 平台的公共图书馆。

1.2.5　Android 平台的开发工具

开发 Android 平台下的应用程序，需要安装以下的工具：

（1）Android SDK：相当于用户在 Java 使用的 JDK。

（2）Eclipse+ADT：安装 Eclipse 开发工具，并为其安装 ADT 插件。

第 2 章 Android 程序设计基础

▲ 引言

Android 平台下的应用程序的结构具有一定规律性，开发者在编写应用程序时，需要了解 Android 应用程序的工作特性，并依据 Android 应用程序设定的目录结构进行程序规范，才能开发出标准高效的应用程序。本章内容为 Android 应用程序开发前准备工作，主要包括开发环境的搭建、Android 应用程序的目录结构、Android 应用程序的生命周期等 Android 程序设计的基础知识。

2.1 Android 开发环境搭建

2.1.1 软件准备

1. 操作系统要求

Android 应用程序可以使用 Windows XP 及其以上版本、MAC OS、Linux 等操作系统进行开发。

> 提 示
>
> 由于不同用户使用的操作系统不同，读者在下载安装开发工具时需选择与系统一致的软件版本。

2. 开发环境要求

Android 应用程序基于 Java 语言进行开发，因此完成开发环境配置时需要进行如下的步骤准备。

（1）构建 Java 程序开发环境（JDK）：JDK 是整个 Java 的核心，包括了 Java 运行环境、Java 工具和 Java 基础类库。

（2）配置 Android 专属的软件开发工具包（Android SDK）：Android SDK 不仅包括了 Android 模拟器，而且包括了各种用来调试、打包和在模拟器上安装应用的工具。

（3）安装 Eclipse 并集成 ADT 插件：Eclipse 是一个开放源代码的、基于 Java 的可扩展开发环境（IDE）。Android 开发者工具(ADT)是 Google 官方所提供的基于 Eclipse 的 Android 开发插件。

以上开发工具均可通过网络下载获得，具体下载地址如表 2-1 所示。

表 2-1　　　　　　　　　　　Android 开发环境所需软件的下载地址表

JDK	http://www.oracle.com/technetwork/java/javase/downloads/index.html
Android SDK	http://developer.android.com/sdk/index.html
Eclipse	http://www.eclipse.org/downloads/
ADT	https://dl-ssl.google.com/android/eclipse/ (该插件需要在 Eclipse 中安装)

> **提　示**
>
> 以上工具不断进行更新完善，因此读者在配置开发环境时下载使用的软件版本信息可能与本书中介绍的版本信息不一致，读者可根据实际情况选择适当的版本信息。

2.1.2　搭建开发环境（WINDOWS 8 64b）

1. 构建 Java 程序开发环境（JDK）

（1）安装 JDK。下载获得 JDK 的安装文件，双击此安装文件打开应用程序开发向导，设置 JDK 的安装目录，如图 2-1 所示。

图 2-1　配置 JDK 的安装目录（一）

单击下一步，执行 JDK 安装，设置 JRE 的安装目录，如图 2-2 所示，单击下一步，执行 JRE 安装，完成整个软件的安装。

（2）配置 JDK 环境变量。右击我的电脑，选择属性，选择高级系统设置，在高级选项卡中选择环境变量，在系统环境变量中，单击新建，设置变量名为 JAVA_HOME，变量值为 JDK 的安装目录，如图 2-3 所示。

（3）检验 Java 开发环境。使用组合键 win+r，在打开中输入 cmd，在打开的命令提示符中输入命令：

java-version，如果显示了当前系统下的 Java 相关版本信息，说明 JDK 安装成功，如图 2-4 所示。

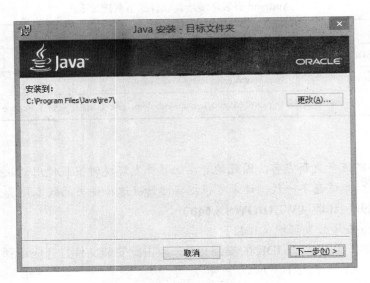

图 2-2 配置 JRK 的安装目录（二）

图 2-3 系统环境变量的配置

2. 配置 Android 专属的软件开发工具包（Android SDK）

（1）安装 Android SDK。双击安装下载获得的 Android SDK 安装文件，单击 Next，会检测当前系统中的 JDK 环境，单击 Next，会要求选择 Android SDK 用户使用权限，配置后单击 Next，设置 Android SDK 的安装目录，如图 2-5 所示，单击 Next 会在 Windows 开始菜单中创建 Android SDK 的快捷方式，设置后单击 Install 后就可以在系统中开始安装 Android

SDK。

图 2-4 Java 开发环境验证

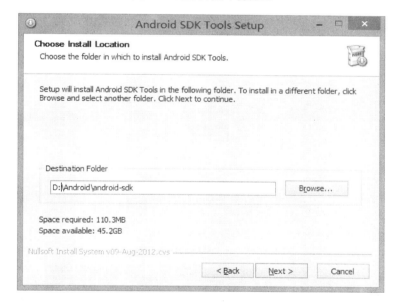

图 2-5 设置 Android SDK 的安装目录

（2）更新 Android SDK。双击运行 Android SDK 安装目录下的 SDK Manager，会打开 Android SDK 管理工具，该工具会自动连网检测 Android SDK 的最新版本和相关最新 Android 的 API 工具，如图 2-6 所示。

根据需要选择要更新的信息，单击 Install，在安装包选择中选择 Accept All 后，系统会自动进行在线更新和相关 Android 包的安装，如图 2-7 所示。

（3）检验 Android SDK。双击 Android SDK 安装目录下的 AVD Manager.exe，会启动 Android 模拟器管理工具平台，如图 2-8 所示。

单击 New 时，可创建不同版本的 Android 模拟器，如图 2-9 所示，读者只需要进行设置模拟器名称并选择模拟器的 Android 版本信息后，单击 Create AVD 就可以创建一个新的 Android 模拟器。

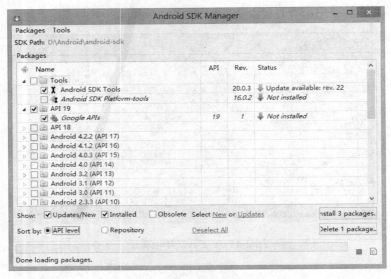

图 2-6　Android SDK 管理界面

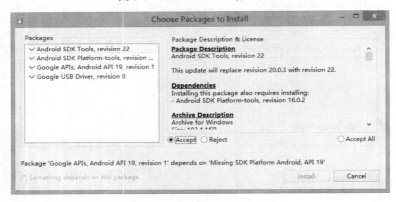

图 2-7　Android SDK 更新选择

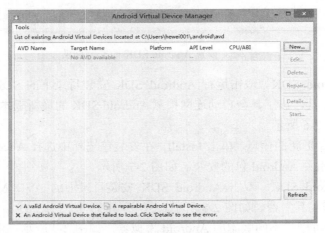

图 2-8　Android 模拟器管理平台

第 2 章　Android 程序设计基础　　15

图 2-9　新的模拟器设置

选择创建成功的模拟器，单击 Start，就可在计算机上运行对应版本的 Android 模拟器，如图 2-10 所示。

图 2-10　Android 模拟器运行效果

3. 安装 Eclipse 并集成 ADT 插件

（1）安装 Eclipse。Eclipse 为一款绿色软件，只需要将下载的压缩包，解压缩到指定的位置就可以完成 Eclipse 的安装。

（2）安装 ADT 插件。为了能够在 Eclipse 下完成 Android 应用程序的开发，同时使用 Android SDK 来运行并验证开发的应用程序，需要为 Eclipse 安装对应的 ADT 插件。双击 Eclipse 开发工具，选择 Help 选项下的 Install new Software，选择 ADD，分别设置插件的名字和对应的下载地址，如图 2-11 所示。

图 2-11　ADT 下载配置信息

选择配置好的插件选项，Eclipse 会自动搜索对应的插件，如图 2-12 所示。

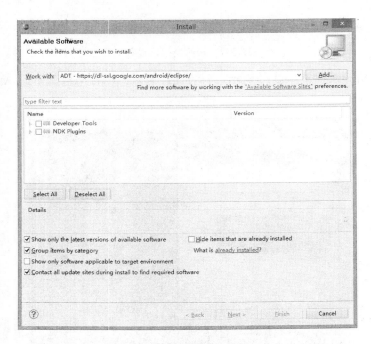

图 2-12　ADT 插件信息

选择需要的插件后，单击 Next 就可以完成插件的安装工作，安装成功后会在 Eclipse 中出现 Android SDK 管理工具和 Android 模拟器管理工具对应的图标。

2.1.3　环境检验-HELLO WORLD

（1）创建虚拟机。打开 Eclipse，选择 Opens the Android Virtual Device Manager，创建模拟器版本，如图 2-13 所示。

第 2 章 Android 程序设计基础

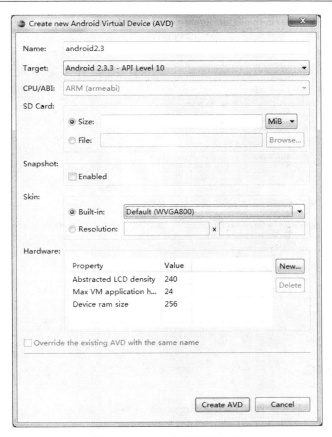

图 2-13 Android 模拟器创建

（2）建立 Android 软件项目。新建工程，选择 Android 下的 Android Application Project，如图 2-14 所示。

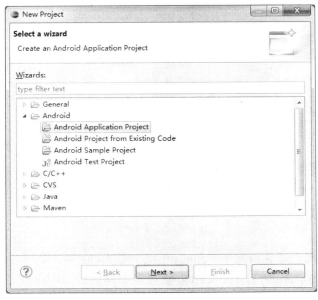

图 2-14 Android 应用程序工程

单击 Next，设置应用程序名、工程名、包名，并选择基于的 Android SDK 版本，如图 2-15 所示。

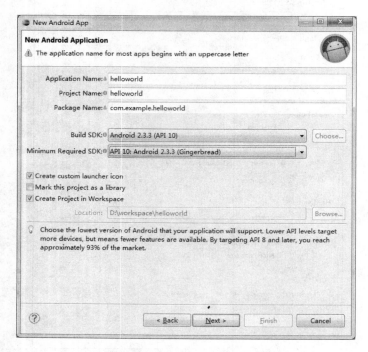

图 2-15　New Android 应用程序设置

单击 Next 设置程序配置文件、应用程序图标、默认界面等信息，单击 Finish，Eclipse 会自动的完成应用程序配置，生成对应的工程项目，如图 2-16 所示。

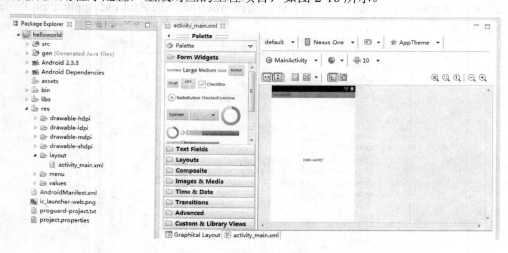

图 2-16　Android 应用工程

（3）运行验证结果。默认情况下 Eclipse 生成一个 Hello World 应用程序，只需右击工程项目，在 Run As 中选择 Android Application 就可以编译运行该应用程序，并在指定的模拟器下面显示运行效果，如图 2-17 所示。

第 2 章 Android 程序设计基础

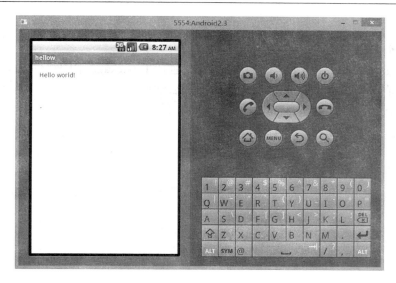

图 2-17 Android 应用程序运行效果

2.2 Android 应用程序目录结构

2.2.1 Android 应用程序的目录结构

当创建 Android 应用工程时，Eclipse 会根据创建的 Android 工程依赖的 Android 系统版本自行创建 Android 应用程序的基本目录，如图 2-18 所示。

Android 应用程序的各个目录被定义了具体的功能，开发人员必须根据不同目录的作用完成代码编写和资源添加，各个目录具体功能如下。

（1）src：存放用户编写的 Java 源代码。

（2）gen：该目录下的文件是不可修改的，是 Eclipse 自动生成的程序。

（3）Android 2.3.3 和 Android Dependencies：对应用程序支持的环境。

（4）assets：放置资源文件。

（5）bin：应用程序生成的文件（运行后的）。

（6）libs：应用程序使用的类库。

（7）res：放置资源文件，不同的资源会在 gen 中生成相应的 id 值，其中：

1）drawable：图片资源。

2）layout：界面布局文件，完成程序的界面设计。

3）menu：主菜单，通过按钮 menu 可显示程序的主菜单。

4）values：程序应用中需要用到的常数值。

（8）AndroidManifest.xml：对整个应用程序进行相关的

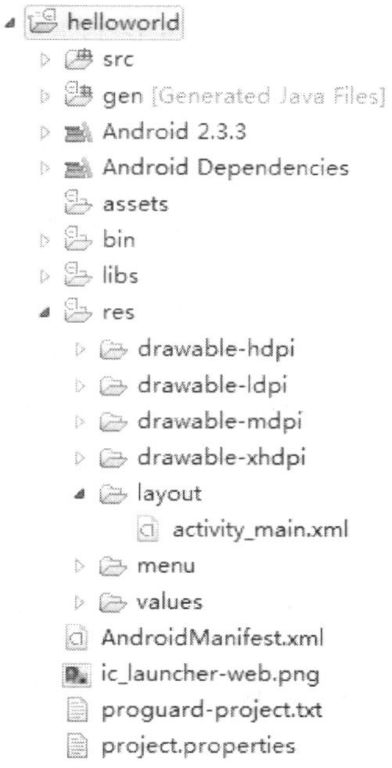

图 2-18 Android 应用程序的基本目录

配置。

2.2.2 AndroidManifest.xml 分析

AndroidManifest.xml 描述了包中的全局数据，包括了 package 中的组件（activities、services 等），它们各自的实现类，各种能被处理的数据和启动位置。

下面是一个简单的 AndroidManifest.xml 文件中的源代码：

```xml
<?xml version="1.0" encoding="utf-8"?>
<manifest xmlns:android="http://schemas.android.com/apk/res/android"
    package="com.example.helloworld"                    //包名
    android:versionCode="1"
    android:versionName="1.0" >                         //应用程序版本信息

    <uses-sdk    //使用的 SDK 版本
        android:minSdkVersion="10"
        android:targetSdkVersion="10" />

    <application
        android:allowBackup="true"
        android:icon="@drawable/ic_launcher"             //引用的图标
        android:label="@string/app_name"                 //引用的应用程序名字
        android:theme="@style/AppTheme" >                //引用的系统样式
        <activity
            android:name="com.example.helloworld.MainActivity"//访问的程序类名
            android:label="@string/app_name" >           //引用的应用程序标签名
            <intent-filter>
                <action android:name="android.intent.action.MAIN" />
      //是程序的入口，决定应用程序最先启动的 Activity
                <category android:name="android.intent.category.LAUNCHER" />
      //决定应用程序是否显示在程序列表里
            </intent-filter>
        </activity>
    </application>

</manifest>
```

在 AndroidManifest.xml 文件中能够声明组件信息（Activities，Content Providers，Services，Intent Receivers），安全控制（permissions），测试（instrumentation），所有声明采用标签的形式进行标识，如表 2-2 所示。

表 2-2　　　　　　　　　　　AndroidManifest.xml 常用标签的作用

标签	作　　用	位置
manifest	描述了包中所有的内容	根节点
uses-permission	应用程序正常运作所需赋予的安全许可	在 manifest 下
permission	声明的安全许可用来限制其他程序对用户包中的组件和功能的使用	在 manifest 下
instrumentation	声明用来测试此包或其他包指令组件的代码	在 manifest 下
application	包含包中应用程序级别组件声明的根节点	在 manifest 下
activity	声明 activity 组件，每一个 activity 必须有一个 <activity> 标记对应	在 application 下

标签	作　　用	位置
receiver	声明 Android 的广播接收器，使得应用程序获得数据的改变或者发生的操作	在 application 下
service	声明 Service，使其能在后台运行的组件	在 application 下
provider	声明 provider，用来管理持久化数据并发布给其他应用程序使用	在 application 下

2.2.3　Android 应用程序的开发过程

Android 应用程序的开发，与用户在多层结构下进行程序开发的步骤相似，用户分层完成不同内容的处理，最终整合成应用程序。在开发过程中，用户主要对 Layout 目录下文件、SRC 目录下文件、Androidmanifest.xml 文件信息进行设定，其中：

（1）Layout 目录：完成界面的设置和布局。

（2）SRC 目录：完成事件的处理。

（3）Androidmanifest.xml：完成对应用程序的整体配置。

一个简单 Android 应用程序的开发过程，如图 2-19 所示。

（1）应用程序资源设置：在 res\drawable 目录下设置应用程序的资源，res\values 文件夹进行键值设置。

（2）gen 目录下的 R.java 会根据 res 目录下的资源自动为每个资源分配 id，方便应用程序调用相应的资源。

（3）在 res\layout 目录下完成不同界面的布局配置。

（4）在 src 目录下完成应用程序的事件处理。

（5）最终在 Androidmanifest.xml 文件中完成程序的总体配置。

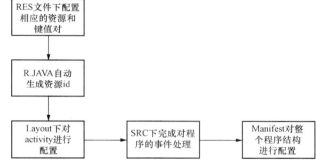

图 2-19　Android 应用程序的开发过程

应用举例 Example_2_1——创建一个 Android 应用程序，分别通过两个按钮实现图片的显示和隐藏，具体步骤如下：

（1）应用程序资源设置：将选定的图片资源文件复制到 res/drawable 文件夹中，修改 res/values/string.xml 中的应用程序全局信息，代码如下：

```
<?xml version="1.0" encoding="utf-8"?>
<resources>
    <string name="app_name">Example_2_1</string>        //应用程序名
    <string name="action_settings">Settings</string>    //主菜单名
</resources>
```

※知识点 1：在 res 下有多个 drawable 前缀的文件夹，分别为 drawable-ldpi（低分辨率）、drawable-mdpi（中分辨率）、drawable-hdpi（高分辨率）、drawable-xhdpi（超高分辨率），用户可根据要映射到模拟器的分辨率进行图片设置。

（2）完成程序的界面设计：本程序用 ImageView 控件存储图片，利用 Button 控件做事件处理控件，拖拽控件面板中 Images&Media 下的 ImageView 控件到程序界面中，并选择已经复制在 res/drawable 中的图片资源，拖拽两个 Button 控件到程序界面中，右击已经在程序界面中的 Button 控件，选择 EditText 设置按钮上显示的文字，本程序设置的主界面效果，如图 2-20 所示。

图 2-20　Example_2_1 界面效果

> **提　示**
>
> 所有的控件效果都可以通过界面对应的 xml 代码文件进行直接设置，如下代码就是本界面对应的 xml 源代码：

```xml
<RelativeLayout xmlns:android="http://schemas.android.com/apk/res/android"
    xmlns:tools="http://schemas.android.com/tools"
    android:layout_width="match_parent"
    android:layout_height="match_parent"
    android:paddingBottom="@dimen/activity_vertical_margin"
    android:paddingLeft="@dimen/activity_horizontal_margin"
    android:paddingRight="@dimen/activity_horizontal_margin"
    android:paddingTop="@dimen/activity_vertical_margin"
    tools:context=".MainActivity" >
    <ImageView
        android:id="@+id/imageView1"                       //控件的 id 名
        android:layout_width="wrap_content"
        android:layout_height="wrap_content"
        android:layout_alignParentTop="true"
        android:layout_centerHorizontal="true"
        android:src="@drawable/android" />
    <Button
        android:id="@+id/button1"
        android:layout_width="wrap_content"
        android:layout_height="wrap_content"
        android:layout_alignLeft="@+id/imageView1"
        android:layout_below="@+id/imageView1"
        android:layout_marginLeft="77dp"
```

```
                android:layout_marginTop="27dp"
                android:text="显示" />                          //控件的现实文本
            <Button
                android:id="@+id/button2"
                android:layout_width="wrap_content"
                android:layout_height="wrap_content"
                android:layout_alignBaseline="@+id/button1"
                android:layout_alignBottom="@+id/button1"
                android:layout_marginLeft="37dp"
                android:layout_toRightOf="@+id/button1"
                android:text="隐藏" />
        </RelativeLayout>
```

※知识点 2：Android 为用户提供了大量可视化空间。开发人员只需要进行简单的拖拽就可以完成应用程序的界面布局。

（3）在 src 目录下完成应用程序的事件处理，本程序源代码如下：

※知识点 3：在完成应用程序的事件处理时，一般要经过映射控件、控件处理两个操作过程。Android 一般通过 findViewById 方法通过界面中控件 ID 来映射显示界面中的控件。

```java
public class MainActivity extends Activity {
  private Button bt1,bt2;
  private ImageView iv;
   public void onCreate(Bundle savedInstanceState) {
        super.onCreate(savedInstanceState);
        setContentView(R.layout.activity_main);
        bt1=(Button)findViewById(R.id.button1); //通过控件的 id 获得界面中的控件
        bt2=(Button)findViewById(R.id.button2);
        iv=(ImageView)findViewById(R.id.imageView1);
        bt1.setOnClickListener(new list1());      //设置按钮的监听器,事件触发时执行
        bt2.setOnClickListener(new list2());
   }
   class list1 implements OnClickListener
   {
     public void onClick(View v) {
        // 当显示按钮触发时,实现图片控件对应的显示
        iv.setVisibility(View.VISIBLE);
     }
   }
   class list2 implements OnClickListener
   {
     public void onClick(View v) {
        // 当隐藏按钮触发时,实现图片控件对应的隐藏
        iv.setVisibility(View.INVISIBLE);
     }
   }
}
```

（4）Androidmanifest.xml 文件。本程序中不需要声明新的组件和权限设置，直接采用 Eclipse 默认方式即可。

2.3 Android 生命周期

2.3.1 Android 多界面切换

一般的应用程序都包含多个界面,会根据不同的用户请求进行页面间的跳转,在 Android 应用程序中通过 Intent 组件实现界面的跳转和数据信息的传递,本节会介绍不同界面间跳转的基本方式,详细的 Intent 使用方式将在第 4 章进行详细介绍。

Android 通过使用 Intent 对象实现多个界面间的切换。具体跳转规则和使用方法如下。

界面 1:(发起)

(1) 创建 Intent 对象;

(2) 设置传递的信息,使用 putExtra(键值名,键值)方法;

(3) 跳转的规则,使用 setClass(界面 1 类,界面 2 类)方法;

(4) 跳转,使用 startActivity(Intent 对象)。

界面 2:(接收)

(5) 获得 Intent,使用 getIntent( )方法;

(6) 读取并使用信息,使用 getExtra(键值名)方法。

应用举例 Example_2_2——在界面 1 设置按钮,单击跳转界面 2。

创建应用程序,分别创建 First、Second 两个 Activity,在界面 1 中添加 Button 控件,在界面 2 中添加 TextView 控件,并设置控件中的文本信息。在不同界面对应的 src 文件中编写如下代码:

First(发起跳转界面)代码如下:

```
public class First extends Activity {
  Button bt;
    public void onCreate(Bundle savedInstanceState) {
        super.onCreate(savedInstanceState);
        setContentView(R.layout.activity_first);
        bt=(Button)findViewById(R.id.button1);
        bt.setOnClickListener(new jt());
    }
    class jt implements OnClickListener
    {
        public void onClick(View v) {
         // 实现界面跳转代码
         Intent t=new Intent();
         t.putExtra("aaa", "我是界面 1 上的数据");
         t.setClass(First.this, Second.class);
         startActivity(t);
      }
    }
}
```

Second(跳转后的界面)代码如下:

```
public class Second extends Activity {
   TextView tv;
    public void onCreate(Bundle savedInstanceState) {
        super.onCreate(savedInstanceState);
```

```
        setContentView(R.layout.activity_second);
        Intent ft=getIntent();
        String faaa=ft.getStringExtra("aaa");
        tv=(TextView)findViewById(R.id.textView1);
        tv.setText(faaa);
    }
}
```

提 示

配置应用程序主界面时需要在Androidmanifest.xml文件中,对应的activity标签声明中追加如下代码:

```
<intent-filter>
    <action android:name="android.intent.action.MAIN" />
    <category android:name="android.intent.category.LAUNCHER" />
</intent-filter>
```

2.3.2 Android DDMS 工具

DDMS 的全称是 Dalvik Debug Monitor Service,是一款功能强大的 Android 系统开发调试工具,DDMS 为 IDE 和模拟器及真正的 Android 设备架起了一座桥梁。开发人员可以通过 DDMS 看到目标机器上运行的进程/线程状态,可以查看进程的 heap 信息,可以查看 logcat 信息,可以查看进程分配内存情况,可以向目标机发送短信及打电话,可以向 android 开发发送地理位置信息。

具体的添加 DDMS 过滤器的方法:在 Eclipse 的 Windows 选项卡中,选择 Open Perspective 中的 Other 选项,选择 DDMS 工具,如图 2-21 所示。

在 DDMS 工具中的 LogCat 中添加监听过滤器,添加要监听的信息,如图 2-22 所示。例如,在 by Log Tag 中添加 System.out,则在 DDMS 中只会显示执行到 System.out 对应的信息。

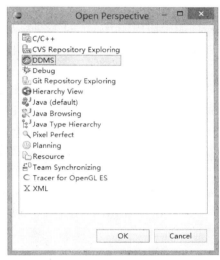

图 2-21 打开 DDMS 工具

图 2-22 添加过滤器的选项卡

※知识点 1：LogCat 中包含 Level、Time、PID、TID、Application、Tag、Text 信息，其中：

（1）Level：表示信息的种类，分为 V（verbose，显示全部信息），D（Debug，显示调试信息），I（Info，显示一般信息），W（Warming，显示警告信息），E（Error，显示错误信息）5 种。

（2）Time：表示执行的时间。

（3）PID：表示程序运行时的进程号。

（4）TID：线程号。

（5）Application：应用程序名字。

（6）Tag：标签，通常表示系统中的一些进程名。

（7）Text：运行时的显示信息。

2.3.3 Android 程序的生命周期

在 Android 应用程序运行过程中，当 Activity 开始启动时，Android 系统会自动调用应用程序的生命周期方法，对于开发人员需要掌握 Android 应用程序运行原理，了解不同生命周期方法的调用过程，Android 应用程序的 Activity 有 7 个生命周期方法。

※知识点 1：生命周期各方法说明

（1）onCreate()方法：当 Activity 第一次创建时调用。

（2）onStart()方法：当 Activity 能被我们看到时调用。

（3）onResume()方法：当 Activity 能获得用户焦点时调用。

（4）onPause()方法：当新的界面挡住原来界面时调用。

（5）onStop()方法：当前界面不可见时要调用这个方法。

（6）onRestart()方法：当重新调用界面时调用。

（7）onDestroy()方法：有两种情况，当调用 activity 的 finish 方法，或者当系统资源不够用的时候。

应用举例 Example_2_3——通过 Example_2_2 案例检测应用程序员多个 Activity 之间跳转时，调用的生命周期方法。

（1）在案例 Example_2_2 基础上复写生命周期方法：在不同界面对应的类中，右击源代码，选择 Source 下的 Override/Implement Methods 选项，在 Activity 中复写 7 个生命周期方法。并在不同方法中添加代码。

First 添加代码：

```
    public void onCreate(Bundle savedInstanceState) {
        // 调用 onCreate 时输出下面的信息
System.out.println("first_____onCreate");
    }
protected void onDestroy() {
      // 调用 onDestroy 时输出下面的信息
     System.out.println("first_____onDestroy");
     super.onDestroy();
   }
    protected void onPause() {
      //调用 onPause 时输出下面的信息
```

```java
        System.out.println("first_____onPause");
        super.onPause();
    }
    protected void onRestart() {
        //调用 onRestart 时输出下面的信息
        System.out.println("first_____onRestart");
        super.onRestart();
    }
    protected void onResume() {
        //调用 onResume 时输出下面的信息
        System.out.println("first_____onResume");
        super.onResume();
    }
    protected void onStart() {
        //调用 onStart 时输出下面的信息
        System.out.println("first_____onStart");
        super.onStart();
    }
    protected void onStop() {
        //调用 onStop 时输出下面的信息
        System.out.println("first_____onStop");
        super.onStop();
    }
```

Second 添加代码：

```java
public class Second extends Activity {
    public void onCreate(Bundle savedInstanceState) {
        //调用 onCreate 时输出下面的信息
        System.out.println("second_____onCreate");
    }
    protected void onDestroy() {
        //调用 onDestroy 时输出下面的信息
        System.out.println("second_____onDestroy");
        super.onDestroy();
    }
    protected void onPause() {
        //调用 onPause 时输出下面的信息
        System.out.println("second_____onPause");
        super.onPause();
    }
    protected void onRestart() {
        //调用 onRestart 时输出下面的信息
        System.out.println("second_____onRestart");
        super.onRestart();
    }
    protected void onResume() {
        //调用 onResume 时输出下面的信息
        System.out.println("second_____onResume");
        super.onResume();
    }
    protected void onStart() {
```

```
        //调用 onStart 时输出下面的信息
        System.out.println("second        onStart");
        super.onStart();
    }
    protected void onStop() {
        //调用 onStop 时输出下面的信息
        System.out.println("second        onStop");
        super.onStop();
    }
}
```

（2）利用 DDMS 监听各生命周期方法：在 LogCat 中添加过滤器，在过滤器的 by Log Tag 中添加 System.out，当不同生命周期运行时，LogCat 就可以监听其中添加的 System.out 输出语句，实现对不同生命周期的监听。

（3）获得监听结果：监听界面创建、界面切换、界面退出时调用的不同生命周期方法。

项目实训 1：简单 Android 应用程序实现

1. 实训目的

（1）掌握 Android 应用程序的基本开发步骤；

（2）熟悉 Eclipse 的基本操作界面，快捷键使用；

（3）掌握 Android 应用程序界面开发的基本方式；

（4）掌握 Android 应用程序代码编写的基本方式；

（5）掌握在 Eclipse 下建立、运行、修改、保存、装入 Android 应用程序。

2. 常见问题分析

（1）使用 Eclipse 编写应用程序时，可以使用 Alt+/快捷键进行代码提示；

（2）编写 Android 应用程序时，通过 setContentView 实现布局文件与后台代码的关联，通过 findViewById 实现控件与代码的关联；

（3）Android 模拟器第一次启动速度很慢，如果需要频繁使用 Android 模拟器建议保持模拟器运行状态，直接编译应用程序即可；

（4）编写应用程序代码时，控件信息需要选择自己建立的程序包中的控件 id，否则会导致应用程序出错，如图 2-23 所示。

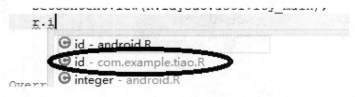

图 2-23　程序包的选择

3. 实训内容

编写 Android 下的应用程序，实现在对话框（EditText）中输入信息，单击按钮（Button）后在屏幕显示信息（TextView），参考效果如图 2-24 所示。

项目实训 2：多界面 Android 应用程序实现

1. 实训目的

（1）掌握 Android 应用程序界面跳转的基本方式；

(2) 熟悉 Eclipse 的基本操作界面，快捷键使用；

(3) 掌握 Android 应用程序的界面跳转时如何进行传值；

(4) 掌握 Android 应用程序 Androidmanifest.xml 基本修改方式。

2. 常见问题分析

(1) 创建多个 Android 界面时，需要在 Androidmanifest.xml 进行每一个界面的声明，同时需要声明主界面。

(2) 创建 Activity 时，给创建的 Activity 命名时需要注意首字母必须大写。

(3) 接收 Intent 传递的数据时，可以用 getStringExtra 接收字符串键值信息，getFloatExtra 接收实型键值对信息，但需要赋初值。

图 2-24　项目实训 1 完成效果图

(4) 出现逻辑错误时，可以通过 DDMS 设置过滤器来协助查找程序中出现的问题。

3. 实训内容

BMI "标准体重计算器" 要求如下：

(1) 如果是男的，则标准体重为（身高–80）×0.7；

(2) 如果是女的，则标准体重为（身高–70）×0.6。

要求：在界面 1 收集用户输入的信息，将收集的信息传给界面 2，在界面 2 显示用户的标准体重信息，参考效果如图 2-25 所示。

图 2-25　项目实训 2 完成效果图

第3章 Android 用户界面开发

▲ 引言

Android 应用程序的基础单元就是 Activity 类中的一个对象。Activity 可以做很多事，比如界面显示、事件处理等。Android 应用程序界面通常使用 View 和 ViewGroup 控件来配合 XML 样式进行设计；而事件则包括按钮事件、触碰事件及一些高级控件的事件监听。下面分别介绍 Activity 的界面设计、显示和事件处理。

3.1 需求分析

3.1.1 七彩文字

实现七彩文字变幻效果，每按一次"点击变色"按钮，文字变一次颜色，在 Android 应用中十分常见。

3.1.2 红心 A 猜测

开发一个小游戏，猜测三张扑克牌，哪个是红心 A，猜对猜错都会有显示，并显示已玩次数和胜率，很好玩，是常见的小游戏。

3.1.3 电子相册

大家都用过 Android 的相册功能，我们就开发一个这样的电子相册，拖曳鼠标可以看到下一张图片，单击鼠标可以弹出图片名称。

3.1.4 游戏登录界面

很多游戏都有登录界面，我们就模拟一个这样的游戏界面。制作一个游戏的菜单，先进入一个用户登录对话框，输入正确的用户名和密码允许登录，然后进入主菜单界面，主菜单界面包括"开始游戏"和"读取进度"两项内容，进入后都会有相应进度条显示。

3.2 知识准备

3.2.1 常用控件使用

1. 文本框（TextView）

作用：显示文本信息。

常用的方法：

（1）setTextColor：改变文本的颜色。

（2）setTextSize：改变文本的大小。
（3）setText：设置显示文字。
（4）setBackgroudColor：设置文字背景颜色。

应用举例 Example_3_1——输出一行文本信息（要求：字体颜色为红色，背景色为绿色，字体大小为 28 号），如图 3-1 所示。

图 3-1 文本框控件的使用

代码如下：

```
public class MainTextView extends Activity {
    private TextView textView;
    public void onCreate(Bundle savedInstanceState) {
        super.onCreate(savedInstanceState);
        setContentView(R.layout.activity_textview);
        //连接 TextView 控件
        textView = (TextView)findViewById(R.id.textView1);
        textView.setTextColor(Color.RED);              //设置字体颜色
        textView.setBackgroundColor(Color.GREEN);      //设置背景颜色
        textView.setTextSize(28);                      //设置字体大小
        textView.setText("使用 textview 的方式");        //显示的文本
    }
}
```

2．提示（Toast）

作用：快速显示讯息（比如短信息提示）。

常用的方法：

`Toast.makeText(this, String,Toast.LENGTH_LONG).show( );`

String：文本信息；

LENGTH_LONG：提示较长时间 ；

show( )：显示提示信息。

应用举例 Example_3_2——使用 Toast 显示一行信息提示（无需设置控件）。

代码：

```
public class MainToast extends Activity {
    public void onCreate(Bundle savedInstanceState) {
        super.onCreate(savedInstanceState);
        setContentView(R.layout.main);
        Toast.makeText(this, "Toast 显示信息", Toast.LENGTH_LONG).show();
    }
}
```

3．按钮（Button）

作用：用户通过按下按钮，或者单击按钮来执行一个动作。

常用的方法：

（1）setTextColor：改变文本的颜色。
（2）setTextSize：改变文本的大小。
（3）setText：设置显示文字。

（4）setBackgroudColor：设置文字背景颜色。
（5）setWidth：设置按钮宽度。
（6）OnClick：处理单击事件。
（7）setOnClickListener：单击事件监听。

应用举例 Example_3_3——设置两个按钮，分别为"显示"、"退出"，单击开始按钮后，设置按钮的宽度、文本的大小、颜色、按钮的背景颜色。单击"显示"按钮后，有一条提示信息；单击"退出"按钮后，退出页面，如图3-2所示。

代码如下：

```java
public class MainActivity extends Activity {
   Button btnToast,btnExit;
    @Override
    public void onCreate(Bundle savedInstanceState) {
        super.onCreate(savedInstanceState);
        setContentView(R.layout.activity_main);
        btnToast = (Button)findViewById(R.id.button1);
        btnExit = (Button)findViewById(R.id.button2);
        //设置按钮宽度
        btnToast.setWidth(200);
        btnExit.setWidth(200);
        //设置按钮文本大小
        btnToast.setTextSize(30);
        btnExit.setTextSize(40);
        //设置按钮文本颜色
        btnToast.setTextColor(Color.BLUE);
        btnExit.setTextColor(Color.RED);
        //设置按钮颜色
        btnToast.setBackgroundColor(Color.YELLOW);
        btnExit.setBackgroundColor(Color.CYAN);
        //给按钮设置监听
        btnToast.setOnClickListener(new toastListener());
        btnExit.setOnClickListener(new exitListener());
   }
    //创建toastListener事件
    class toastListener implements OnClickListener
    {
      public void onClick(View v) {
         Toast.makeText(MainActivity.this, "按钮被按下",Toast.LENGTH_LONG).show();// Toast 提示
      }
    }
    //创建exitListener事件
    class exitListener implements OnClickListener
    {
      public void onClick(View v) {
         MainActivity.this.finish();// 退出
      }
    }
}
```

4. 带图标按钮（ImageButton）

作用：带图标的按钮。通过 SetImageDrawable 方法设置按钮要显示的图标。

方法：同 Button。

应用举例 Example_3_4——添加三个图标按钮，其中一个为系统图标。单击系统图标时，提示信息"使用的是系统图标"；当单击第二个图标按钮时，将图标变为系统图标，如图 3-3 所示。

图 3-2　按钮控件的使用

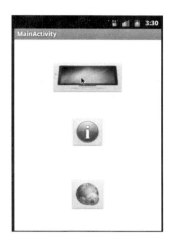

图 3-3　带图标按钮的使用

代码如下：

```
public class MainActivity extends Activity {
    /** Called when the activity is first created. */
    // @Override
   private ImageButton imgbtnSys,imgbtnSencond,imgbtnThird;//声明三个图标按钮
    public void onCreate(Bundle savedInstanceState) {
        super.onCreate(savedInstanceState);
        setContentView(R.layout.activity_main);
        //连接控件
        imgbtnSys = (ImageButton)findViewById(R.id.imageButton1);
        imgbtnSencond = (ImageButton)findViewById(R.id.imageButton2);
        imgbtnThird = (ImageButton)findViewById(R.id.imageButton3);
        //设置监听
        imgbtnSys.setOnClickListener(new sysListener());
        imgbtnSencond.setOnClickListener(new secondListener());
        imgbtnThird.setOnClickListener(new thirdListener());
    }
    //创建系统图标按钮监听事件
    class sysListener implements OnClickListener{
     //@Override
      public void onClick(View arg0) {
        // TODO Auto-generated method stub
        Toast.makeText(MainActivity.this, "使用的是系统图标", Toast.LENGTH_SHORT).show();
      }
    }
    //创建第二个图标按钮监听事件
    class secondListener implements OnClickListener {
```

```
        //@Override
        public void onClick(View arg0) {
            // 更改/设置图标按钮上显示的图片
    imgbtnSencond.setImageDrawable(getResources().getDrawable(R.drawable.button3));
            Toast.makeText(MainActivity.this, "您单击的是第二个按钮！", Toast.LENGTH_SHORT).show();
        }
    }
    //创建第三个图标按钮监听事件
     class thirdListener implements OnClickListener{
       //@Override
        public void onClick(View arg0) {
            // TODO Auto-generated method stub
            Toast.makeText(MainActivity.this, "您单击的是第三个按钮！", Toast.LENGTH_SHORT).show();
        }
    }
}
}
```

5. 编辑框（EditText）

说明：用户输入信息。

常用的方法：

（1）getText( )获得文本框中的内容。

（2）setHint( )初始化文本框中信息。

（3）对文本框输入监听：setOnKeyListener 实现。

应用举例 Example_3_5——用户输入信息后，获得用户输入的内容，显示在一个 TextView 上，如图 3-4 所示。

代码如下：

图 3-4 编辑框控件的使用

```
public class MainActivity extends Activity {
    /** Called when the activity is first created. */

    private EditText edtInput;
    private TextView tvDisplay;
    @Override
    public void onCreate(Bundle savedInstanceState) {
        super.onCreate(savedInstanceState);
        setContentView(R.layout.activity_main);
        //连接控件
        edtInput = (EditText)findViewById(R.id.editText1);
        tvDisplay = (TextView)findViewById(R.id.textView1);
        //设置监听
        edtInput.setOnKeyListener(new keyListener());
    }
    //创建文本框按钮监听事件
    class keyListener implements OnKeyListener
    {
```

```
            public boolean onKey(View v, int keyCode, KeyEvent event) {
                // 在 TextView 上显示 EditText 中输入的文本
                tvDisplay.setText(edtInput.getText());
                return false;
            }
        }
    }
```

6. 单项选择（RadioGroup、RadioButton）

作用：使用 RadioGroup 与 RadioButton 组合实现单项选择的效果。

常用方法：

（1）setOnCheckedChangeListener：监听按钮改变（其中 checkedId 为选择项的标识 ID）。

（2）isChecked( )：是否选择。

（3）getId( )：返回该单选按钮组中所选择的单选按钮的标识 ID，如果没有勾选则返回 –1。

使用步骤：

（1）定义 RadioGroup，并包含若干 RadioButton。

（2）监听 RadioGroup 的 setOnCheckedChangeListener。

（3）比较选项是否一致：通过 getId 获得选择项信息。

（4）反馈结果。

应用举例 Example_3_6——实现单项选择题，Android 底层基于什么操作系统？

（a）Windows；（b）Linux；（c）Mac os；（d）Java

如果回答正确，提示"回答正确"；回答错误提示"回答错误"，如图 3-5 所示。

代码如下：

图 3-5　单项选择

```
public class MainActivity extends Activity {
    /** Called when the activity is first created. */
    private RadioGroup radioGroup;                          //声明单选按钮组
    private RadioButton radioButton;                        //声明单选按钮
    @Override
    public void onCreate(Bundle savedInstanceState) {
        super.onCreate(savedInstanceState);
        setContentView(R.layout.activity_main);
        radioGroup = (RadioGroup)findViewById(R.id.radioGroup1);
        radioButton = (RadioButton)findViewById(R.id.radioButton2);
        radioGroup.setOnCheckedChangeListener(new choiceListener());
                                                            //设置单选按钮组的监听
    }
    class choiceListener implements OnCheckedChangeListener//创建监听事件
    {
        public void onCheckedChanged(RadioGroup group, int checkedId) {
            // 判断选择的选项 ID 是否与正确答案的 ID 相等
            if(checkedId == radioButton.getId())
            {
```

```
            Toast.makeText(MainActivity.this,"回答正确",Toast.LENGTH_LONG).show();
        }
        else
        {
            Toast.makeText(MainActivity.this,"回答错误",Toast.LENGTH_LONG).show();
        }
    }
}
```

7. 多项选择（CheckBox）

作用：实现多项选择功能。

常用方法：

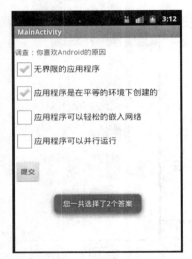

（1）setOnCheckedChangeListener：监听按钮改变。

（2）isChecked( )：是否选择。

使用步骤：

（1）定义 CheckBox。

（2）通过 setOnCheckedChangeListener 监听每一个 CheckBox（可选）。

（3）判断是否选中。

应用举例 Example_3_7——实现多项选择功能，调查：你喜欢 Android 的原因。

(a) 无界限的应用程序；

(b) 应用程序是在平等的环境下创建的；

(c) 应用程序可以轻松的嵌入网络；

(d) 应用程序可以并行运行。

图 3-6 多项选择控件的使用

显示结果为：您一共选择了 n 项，如图 3-6 所示。

代码如下：

```
public class MainActivity extends Activity {
    /** Called when the activity is first created. */
    private Button btnPost;//声明提交按钮
    private CheckBox checkBox1,checkBox2,checkBox3,checkBox4;//声明四个复选框
    @Override
    public void onCreate(Bundle savedInstanceState) {
        super.onCreate(savedInstanceState);
        setContentView(R.layout.activity_main);
        btnPost = (Button)findViewById(R.id.button1);              //连接控件
        checkBox1 = (CheckBox)findViewById(R.id.checkBox1);
        checkBox2 = (CheckBox)findViewById(R.id.checkBox2);
        checkBox3 = (CheckBox)findViewById(R.id.checkBox3);
        checkBox4 = (CheckBox)findViewById(R.id.checkBox4);
        //设置按钮监听
        btnPost.setOnClickListener(new postListener());
    }
    //创建监听事件
```

```
    class postListener implements OnClickListener
    {
      public void onClick(View v) {                    //计算选中几个复选框
        // TODO Auto-generated method stub
        int num = 0;
        if(checkBox1.isChecked())
        {
           num++;
        }
        if(checkBox2.isChecked())
        {
           num++;
        }
        if(checkBox3.isChecked())
        {
           num++;
        }
        if(checkBox4.isChecked())
        {
           num++;
        }
        //Toast 提示信息
        Toast.makeText(MainActivity.this, "您一共选择了" + num + "个答案", Toast.LENGTH_LONG).show();
      }
    }
}
```

8. 图片视图（ImageView）

功能：显示图片。

常用方法：

setImageResource：设置显示的图片资源。

9. 日期（DatePicker）

作用：实现显示设置日期。

常用方法：

（1）getDayOfMonth 获取选择的天数。

（2）getMonth 获取选择的月份。

（3）getYear 获取选择的年份。

（4）init 初始化状态。

（5）setEnabled 设置视图的启用状态。

（6）OnDateChange 日期更改时进行处理。

（7）OnDateChangeListener 日期调整事件监听。

10. 时间（TimePicker）

作用：实现显示设置时间。

常用方法：

（1）is24HourView 获取当前系统设置是否是 24 小时制。

图 3-7 日期和时间控件的使用

（2）setCurrentHour 设置当前小时。
（3）setCurrentMinute 设置当前分钟（0～59）。
（4）setEnabled 设置可用的视图状态。
（5）setIs24HourView 设置是 24 小时还是上午/下午制。
（6）OnTimeChange 时间改变时进行处理。
（7）setOnTimeChangedListener 时间调整事件监听。

应用举例 Example_3_8——修改日期和时间。

两种修改方式：一种在界面上直接修改，并且在第二个文本框上显示修改结果；第二种通过单击按钮，在弹出的对话框中修改，并在第三个文本框中显示修改结果，如图 3-7 所示。

代码如下：

```java
public class MainActivity extends Activity {
 /** Called when the activity is first created. */
// @Override
private DatePicker datePicker;           //声明日期控件
private TimePicker timePicker;           //声明时间控件
private Button btnDate,btnTime;          //声明两个按钮
private TextView tvDate,tvTime;//tvDate 修改后文本框,tvTime 对话框修改后文本框
private int year,month,day,hour,minute;
Calendar calendar;
 public void onCreate(Bundle savedInstanceState) {
    super.onCreate(savedInstanceState);
    setContentView(R.layout.activity_main);
    //连接控件
    datePicker = (DatePicker)findViewById(R.id.datePicker1);
    timePicker = (TimePicker)findViewById(R.id.timePicker1);
    btnDate = (Button)findViewById(R.id.button1);
    btnTime = (Button)findViewById(R.id.button2);
    tvDate = (TextView)findViewById(R.id.textView2);
    tvTime = (TextView)findViewById(R.id.textView3);
    //初识化 Calendar
    calendar = Calendar.getInstance();
    //获取当前时间、日期
    year = calendar.get(Calendar.YEAR);
    month = calendar.get(Calendar.MONTH);
    day = calendar.get(Calendar.DAY_OF_MONTH);
    hour = calendar.get(Calendar.HOUR);
    minute = calendar.get(Calendar.MINUTE);
    //设置成 24 小时制
    timePicker.setIs24HourView(true);
    //给控件设置监听
    //直接在控件上修改
    datePicker.init(year, month, day, new datechangejt1());
    timePicker.setOnTimeChangedListener(new timechangejt1());
    //在对话框的控件上修改
    btnDate.setOnClickListener(new datechangejt2());
```

```java
            btnTime.setOnClickListener(new timechangejt2());
    }
    //给设置日期按钮设置监听事件
    class datechangejt2 implements OnClickListener {
      //@Override
      public void onClick(View v) {
        // TODO Auto-generated method stub
        new DatePickerDialog(MainActivity.this, new jt3(), year, month, day).show();
      }
       class jt3 implements OnDateSetListener {
         //@Override
         public void onDateSet(DatePicker view, int year, int monthOfYear,
             int dayOfMonth) {
           // TODO Auto-generated method stub
           year = year;
           month = monthOfYear;
           day = dayOfMonth;
           updatedisplay2();
         }
       }
    }
//给设置时间按钮设置监听事件
    class timechangejt2 implements OnClickListener {
      //@Override
      public void onClick(View v) {
        // TODO Auto-generated method stub
      new TimePickerDialog(MainActivity.this, new jt4(), hour, minute, true).show();
      }
       class jt4 implements OnTimeSetListener {
         public void onTimeSet(TimePicker view, int hourOfDay, int minute) {
           // TODO Auto-generated method stub
           hour = hourOfDay;
           minute = minute;
           updatedisplay2();
         }
       }
    }
    class datechangejt1 implements OnDateChangedListener {
      //@Override
      public void onDateChanged(DatePicker view, int year, int monthOfYear,
          int dayOfMonth) {
        // TODO Auto-generated method stub
        //设置日期代码
        year = year;
        month = monthOfYear;
        day = dayOfMonth;
        updatedisplay1();//在 TextView 上显示
      }
    }
    class timechangejt1 implements OnTimeChangedListener {
      //@Override
```

```java
        public void onTimeChanged(TimePicker view, int hourOfDay, int minute) {
            // TODO Auto-generated method stub
            //设置时间
            hour = hourOfDay;
            minute = minute;
            updatedisplay1();//在TextView上显示
        }
    }
    private void updatedisplay1()
    {
        StringBuilder s = new StringBuilder().append(year).append("/").append(month+1).append("/").append(day).append("  ").append(format(hour)).append(":").append(format(minute));
        tvDate.setText(s);
    }
    private void updatedisplay2()
    {
        StringBuilder s = new StringBuilder().append(year).append("/").append(month+1).append("/").append(day).append("  ").append(format(hour)).append(":").append(format(minute));
        tvTime.setText(s);
    }
    private String format(int x)
    {
        String s = "" + x;
        if(s.length() == 1)
        {
            s = "0" + s;
        }
        return s;
    }
}
```

3.2.2 绑定类控件

绑定类控件包括列表(ListView)、下拉菜单(Spinner)、自动提示(AutoCompleteTextView)、拖动效果（Gallery）、网格视图(GridView) 等类型，这些控件可以设计不同行绑定不同数据。

1. 数据与控件动态绑定步骤

（1）准备数据。

（2）数据映射到适配器(ArrayAdapter、BaseAdapter、 SimpleCursorAdapter)。

（3）将适配器与控件绑定 setAdapter。

2. 适配器

Android 的 Adapter 是连接后端数据和前端显示的适配器接口，如图 3-8 所示。

实现数据的主要适配器如下：

ArrayAdapter：最简单，对应数组，只能展示一行字。

SimpleCursorAdapter：对数据库的数据。

BaseAdapter：性能最强，难度最大（最实用、最高科技）。

第 3 章 Android 用户界面开发

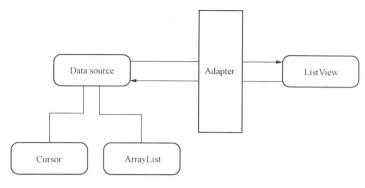

图 3-8 适配器结构图

3. ArrayAdapter

适用于纯文本类型数据，应用于列表(ListView)、下拉菜单（Spinner）、自动提示（AutoCompleteTextView），如图 3-9 所示。

图 3-9 ArrayAdapter 应用的三种类型

ArrayAdapter 用法：

（1）方式：ArrayAdapter（ Context context， int layout，数据）。

（2）Context context：访问资源（一般使用 this 来访问）。

（3）int layout：系统定义好的样式。

（4）数据：一般是一个一维数组。

应用举例 Example_3_9——将周一到周日选项绑定到 ListView 上，并触发滚动和单击事件。结果在 TextView 上显示选择结果，如图 3-10 所示。

代码如下：

图 3-10 ListView 控件的使用

```
public class listview extends Activity {
   /** Called when the activity is first created. */
   private String[] array=
   {
       "星期一","星期二","星期三","星期四","星期五","星期六","星期日"
   };
   private ListView lv;
   private TextView tv;
```

```java
    @Override
    public void onCreate(Bundle savedInstanceState) {
        super.onCreate(savedInstanceState);
        setContentView(R.layout.main);

        tv=(TextView)findViewById(R.id.textView1);
        lv=(ListView)findViewById(R.id.listView1);
        lv.setBackgroundColor(Color.GREEN);
        //建立适配器
        ArrayAdapter<String> aa=new ArrayAdapter<String>(this,android.R.layout.select_dialog_singlechoice,array);
        //ListAdapter aa=new SimpleCursorAdapter(this, 0, null, array, null);
        //将适配器添加到ListView中
        lv.setAdapter(aa);
        lv.setOnItemClickListener(new jt());
        lv.setOnItemSelectedListener(new jt1());
    }
    class jt implements OnItemClickListener
    {

        @Override
        public void onItemClick(AdapterView<?> arg0, View arg1, int arg2,
            long arg3) {
          // TODO Auto-generated method stub
          tv.setText("你选择的是第:"+arg2+"项");
        }

    }
    class jt1 implements OnItemSelectedListener
    {

        @Override
        public void onItemSelected(AdapterView<?> arg0, View arg1, int arg2,
            long arg3) {
          // TODO Auto-generated method stub
          tv.setText("你选择的是第:"+arg2+"项");
        }

        @Override
        public void onNothingSelected(AdapterView<?> arg0) {
            // TODO Auto-generated method stub

        }

    }
}
```

应用举例 Example_3_10——spinner 使用下拉菜单完成血型选择，如图 3-11 所示。
代码如下：

```java
public class spinner extends Activity {
    /** Called when the activity is first created. */
```

```
    private Spinner sp;
    private static String[] aa=new String[]
       {
         "O型","A型","B型","AB型","其他"
       };
    private TextView tv;
     @Override
     public void onCreate(Bundle savedInstanceState) {
        super.onCreate(savedInstanceState);
        setContentView(R.layout.main);
        sp=(Spinner)findViewById(R.id.spinner1);

        ArrayAdapter<String> ap=new ArrayAdapter<String>(this,android.R.layout.select_dialog_multichoice,aa);ap.setDropDownViewResource(android.R.layout.select_dialog_singlechoice);
        sp.setAdapter(ap);
        sp.setPrompt("请选择血型");
     }
}
```

应用举例 Example_3_11——AutoCompleteTextView 输入信息实现自动信息提示功能，如图 3-12 所示。

图 3-11 spinner 控件的使用　　　图 3-12 AutoCompleteTextView 控件的使用

代码如下：

```
public class auto extends Activity {
   /** Called when the activity is first created. */
   private AutoCompleteTextView act;
   private static String[] str=new String[]
{
"aaaaa","bbbbb","ccccc","aaabb","aaacc","aaadd","aabbb","aaccc","bbaaa","bbccc"
};
   @Override
   public void onCreate(Bundle savedInstanceState) {
```

```
        super.onCreate(savedInstanceState);
        setContentView(R.layout.main);act=(AutoCompleteTextView)findViewById
(R.id.autoCompleteTextView1);
        ArrayAdapter<String> aa=new ArrayAdapter<String>(this,android.R.layout.
select_dialog_item,str);
        act.setAdapter(aa);
    }
}
```

4. 重写 BaseAdapter

（1）新创建一个类继承 BaseAdapter。

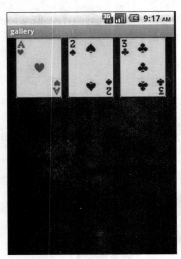

（2）重写 BaseAdapter 的各个方法。
（3）创建新的类的构造函数。
（4）对各个方法进行实现。

具体实现方法如下：

创建类——extends BaseAdapter——添加相关方法——创建构造函数 public imageadapter (Context c)。

注：
Gallery 常用方法：
（1）OnItemClickListener：所选项被单击事件。
（2）OnItemSelectedListener：所选项被改变事件。
GridView 常用方法：
OnItemClickListener：所选项被单击事件。

应用举例 Example_3_12——用 gallery 使用无敌模式，如图 3-13 所示。

图 3-13 用 gallery 使用无敌模式

代码如下：

新建的类：

```java
public class imageadapter extends BaseAdapter {
   Context mc;
   private int[] id=
   {
        R.drawable.p01,
        R.drawable.p02,
        R.drawable.p03,
        R.drawable.p04
   };
    public imageadapter(Context c)
    {
        mc=c;
    }
    public int getCount() {
       return id.length;
    }
    public Object getItem(int position) {
      return position;
```

```
        }
    public long getItemId(int position) {
        return position;
    }
    public View getView(int position, View convertView, ViewGroup parent) {
        ImageView iv=new ImageView(mc);
        iv.setImageResource(id[position]);
        iv.setLayoutParams(new Gallery.LayoutParams(120,120));
        iv.setScaleType(ImageView.ScaleType.FIT_CENTER);
        return iv;
    }
}
public class gallery extends Activity {
    private Gallery ge;
        public void onCreate(Bundle savedInstanceState) {
        super.onCreate(savedInstanceState);
        setContentView(R.layout.main);
        ge=(Gallery)findViewById(R.id.gallery1);
        ge.setAdapter(new imageadapter(this));
        ge.setBackgroundColor(R.drawable.zbj);
        ge.setOnItemClickListener(new jt());
    }
    class jt implements OnItemClickListener
    {
      public void onItemClick(AdapterView<?> arg0, View arg1, int arg2,
            long arg3) {
            Toast.makeText(gallery.this, "你选择了"+arg2, Toast.LENGTH_LONG).
show();
        }
    }
}
```

应用举例 Example_3_13——用 girdview 使用无敌模式，如图 3-14 所示。

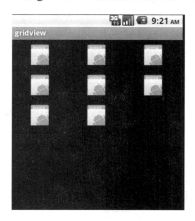

图 3-14 用 girdview 使用无敌模式

代码如下：

```
public class gridadapter extends BaseAdapter{
```

```java
        Context mc;
        int[] ma;
        public gridadapter(Context c,int [] a)
        {
           mc=c;
           ma=a;
        }
        public int getCount() {
           return ma.length;
        }
        public Object getItem(int position) {
           return position;
        }
        public long getItemId(int position) {
           return position;
        }
        public View getView(int position, View convertView, ViewGroup parent) {
           ImageView iv=new ImageView(mc);
           iv.setImageResource(ma[position]);
           return iv;
        }
}
public class grid extends Activity {
        private GridView gv;
    private int[] tp=new int[]
{
    R.drawable.icon,R.drawable.icon,R.drawable.icon,R.drawable.icon,R.drawable.icon,R.drawable.icon,R.drawable.icon,R.drawable.icon
        };
    public void onCreate(Bundle savedInstanceState) {
        super.onCreate(savedInstanceState);
        setContentView(R.layout.main);
        gv=(GridView)findViewById(R.id.gridView1);
        gridadapter gd=new gridadapter(this,tp);
        gv.setAdapter(gd);
        gv.setOnItemClickListener(new jt());
    }
    class jt implements OnItemClickListener
    {
      public void onItemClick(AdapterView<?> arg0, View arg1, int arg2,
           long arg3) {
        Toast.makeText(grid.this, "你显示的是"+arg2, Toast.LENGTH_SHORT).show();
      }
    }
}
```

应用举例 Example_3_14——用 baseadapter 绑定布局 listview（文字+图片模式）使用无敌模式，如图 3-15 所示。

代码如下：

定义背景的 xml 文件（包含 textview imageview：LinearLayout）。

```java
public class base extends BaseAdapter{
    Context mc;
    String[] ms;
    int[] mid;
    public base(Context c,String[] s,int[] id)
    {
        mc=c;
        ms=s;
        mid=id;
    }
    public int getCount() {
        // TODO Auto-generated method stub
        return ms.length;
    }
    public Object getItem(int position) {
        return position;
    }
    public long getItemId(int position) {
        return position;
    }
    public View getView(int position, View convertView, ViewGroup parent) {
        LayoutInflater li=LayoutInflater.from(mc);
        View v=li.inflate(R.layout.bj, null);
        ImageView iv=(ImageView)v.findViewById(R.id.imageView1);
        TextView tv=(TextView)v.findViewById(R.id.textView1);
        iv.setImageResource(mid[position]);
        tv.setText(ms[position]);
        return v;
    }
}
public class wd extends Activity {
    /** Called when the activity is first created. */
    private ListView lv;
    private String[] name=new String[]
{
        "张三","李四","王五"
};
    private int[] aa=new int[]
                    {
            R.drawable.icon,R.drawable.icon,R.drawable.icon,
                    };
    @Override
    public void onCreate(Bundle savedInstanceState) {
        super.onCreate(savedInstanceState);
        setContentView(R.layout.main);

        lv=(ListView)findViewById(R.id.listView1);
        base bs=new base(this,name,aa);
        lv.setAdapter(bs);

    }
}
```

图 3-15　用 baseadapter 绑定布局

3.2.3 高级控件的使用

1. 菜单（Menu）

步骤：

1）建立菜单界面：android xml file。

2）设置菜单选项：id 和 title。

3）复写 onOptionsItemSelected（菜单事件）。

获得所选项 id 信息，通过比较事件，确定要完成的任务。

常用的方法：

1）getMenuInflater( )：获得菜单界面。

2）inflate（资源，menu）：获取界面布局。

3）getItemId( )：获得所单击时间的 Id。

应用举例 Example_3_15——为主界面增加菜单，实现软件版本说明和软件退出功能，如图 3-16 所示。

图 3-16 菜单设计

代码如下：

```
public class MainActivity extends Activity {
    public void onCreate(Bundle savedInstanceState) {
        super.onCreate(savedInstanceState);
        setContentView(R.layout.activity_main);
    }
    public boolean onCreateOptionsMenu(Menu menu) {
        getMenuInflater().inflate(R.menu.activity_main, menu);
        return true;
    }
    public boolean onOptionsItemSelected(MenuItem item) {
        // TODO Auto-generated method stub
        int i=item.getItemId();
        switch (i) {
        case R.id.menu_settings:
            Intent t=new Intent();
            t.setClass(MainActivity.this, Guanyu.class);
            startActivity(t);
            break;
        case R.id.exit:
            MainActivity.this.finish();
            break;
        }
        return true;
    }
}
```

2. 对话框（Dialog）

（1）AlertDialog:自定义对话框。

步骤：

1）Builder 名=new AlertDialog.Builder(this)。

2）设置相关属性信息。

常用的方法：

setTitle（信息）：给对话框设置标题。

setIcon：给对话框设置图标。

setMessage（信息）：设置对话框提示信息。

setPositiveButton（名，监听器）：添加 Yes 按钮。

setNegativeButton（名，监听器）：添加 No 按钮。

setView( )

Create：创建。

Show：显示。

应用举例 Example_3_16——设计一个简单的对话框，输出如图 3-17 所示的结果。

代码如下：

图 3-17 简单对话框界面

```
public class zhujiemian extends Activity {
    public void onCreate(Bundle savedInstanceState) {
        super.onCreate(savedInstanceState);
        setContentView(R.layout.main);
        Builder b=new AlertDialog.Builder(this);
        b.setTitle("提示信息");
        b.setMessage("这是一个很重要的对话框！");
        b.setPositiveButton("确定", new jt());
        b.create().show();
    }
    class jt implements OnClickListener
    {
      public void onClick(DialogInterface dialog, int which) {
            zhujiemian.this.finish();
        }
    }
}
```

图 3-18 自定义对话框的使用

（2）交互式 AlertDialog 的定义。

步骤：

Builder 名=new AlertDialog.Builder(this)。

view 名=LayoutInflater(). inflate(资源，null)（绘制界面）。

setview 设置绘制成的界面。

属性设置。

应用举例 Example_3_17——在对话框上定义一个 EditText，单击"确定"按钮时，主界面获得这个 EditText 中的内容，如图 3-18 所示。

代码如下：

```
public class zdydlg extends Activity {
    /** Called when the activity is first created. */
```

```java
        private TextView tv;
        private EditText et;
        private View v;
         @Override
         public void onCreate(Bundle savedInstanceState) {
            super.onCreate(savedInstanceState);
            setContentView(R.layout.main);
            tv=(TextView)findViewById(R.id.Text1);
            Builder dlg=new AlertDialog.Builder(zdydlg.this);
            v=getLayoutInflater().inflate(R.layout.cd, null);
            dlg.setView(v);
            dlg.setTitle("发送短消息");
            dlg.setMessage("输入短信内容");
            dlg.setPositiveButton("确定", new jt1());
            dlg.setNegativeButton("取消", new jt2());
            dlg.create().show();
        }
        class jt1 implements OnClickListener
        {
          public void onClick(DialogInterface dialog, int which) {
            // TODO Auto-generated method stub
            et=(EditText)v.findViewById(R.id.editText1);
            tv.setText(et.getText().toString());

          }
        }
        class jt2 implements OnClickListener
        {
          public void onClick(DialogInterface dialog, int which) {
            // TODO Auto-generated method stub
            zdydlg.this.finish();
          }
        }
}
```

3. 进度窗体（ProgressDialog）

作用：当程序需要加载时，弹出一个进度窗体，等待程序执行，加载结束时关闭。

步骤：

1）创建一个 ProgressDialog。

2）设置属性，并且显示。

3）程序执行后注销窗体。

常用的方法：

1）setProgressStyle：设置风格（环形：STYLE_SPINNER，柱形： STYLE_HORIZONTAL）。

2）setTitle：设置题目。

3）setMessage：设置内容。

4）setProgress：设置进度条（柱形时有效）。

5）show：显示窗体。

6）dismiss( )：注销窗体。

应用举例 Example_3_18——单击程序加载程序（10 秒），回到主界面，如图 3-19 所示。代码如下：

```
public class progress extends Activity {
    private Button bt;
  private ProgressDialog pd;
     public void onCreate(Bundle savedInstanceState) {
       super.onCreate(savedInstanceState);
       setContentView(R.layout.main);
       bt=(Button)findViewById(R.id.button1);
       bt.setOnClickListener(new jt());
   }
   class jt implements OnClickListener
   {
     public void onClick(View v) {
       pd=ProgressDialog.show(progress.this, "请稍等片刻", "程序正在运行中……");
       new Thread()
       {
         public void run()
         {
           try {
             sleep(10000);
           } catch (InterruptedException e) {
             e.printStackTrace();
           }
           pd.dismiss();
         }
       }.start();
     }
   }
}
```

应用举例 Example_3_19——单击程序进行安装（带进度提示），安装成功返回主菜单，如图 3-20 所示。

图 3-19　进度窗体的使用　　　　图 3-20　进度条提示窗口

代码如下:

```java
public class progress extends Activity {
    /** Called when the activity is first created. */
    private Button bt;
    private ProgressDialog pd;
    @Override
    public void onCreate(Bundle savedInstanceState) {
        super.onCreate(savedInstanceState);
        setContentView(R.layout.main);
        bt=(Button)findViewById(R.id.button1);

        bt.setOnClickListener(new jt());
    }
    class jt implements OnClickListener
    {
      public void onClick(View v) {
         pd=new ProgressDialog(progress.this);
         pd.setProgressStyle(ProgressDialog.STYLE_HORIZONTAL);
         pd.setTitle("请稍等片刻");
         pd.setMessage("程序正在运行中");
         pd.setProgress(100);
         pd.show();
         new Thread()
         {
           public void run()
           {
              try {
                 int m=0;
                 while(m<=100)
                 {
                    pd.setProgress(m++);
                    sleep(100);
                 }
              } catch (InterruptedException e) {
                 // TODO Auto-generated catch block
                 e.printStackTrace();
              }
              pd.dismiss();
           }
         }.start();
      };
    }
}
```

4. 拖动条（SeekBar）

（1）作用：进行一些状态的调节（比如：音量）。

（2）监听 OnSeekBarChangeListener 事件包含：

1）onProgressChanged：数值改变（0-100）。

2）onStartTrackingTouch：开始拖动。

3) onStopTrackingTouch：停止拖动。

应用举例 Example_3_20——进行声音调节，分别监视调节的值，调节的状态，如图 3-21 所示。

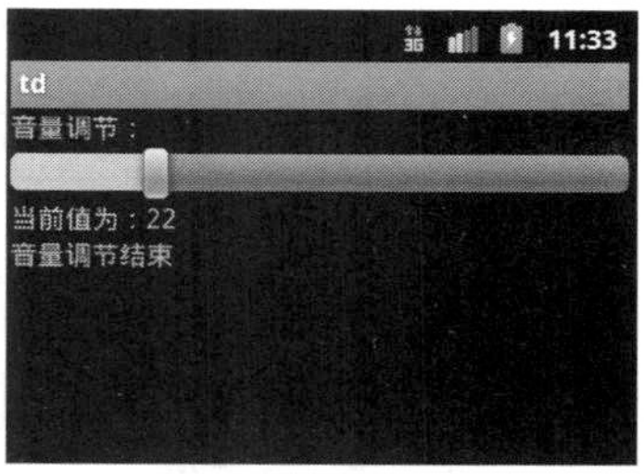

代码如下：

```java
public class td extends Activity {
    /** Called when the activity is first created. */
    private SeekBar seb;
    private TextView tv1,tv2;
    @Override
    public void onCreate(Bundle savedInstanceState) {
        super.onCreate(savedInstanceState);
        setContentView(R.layout.main);
        seb=(SeekBar)findViewById(R.id.seekBar1);
        tv1=(TextView)findViewById(R.id.textView1);
        tv2=(TextView)findViewById(R.id.textView2);
        seb.setOnSeekBarChangeListener(new jt());
    }
    class jt implements OnSeekBarChangeListener
    {
      public void onProgressChanged(SeekBar seekBar, int progress,
          boolean fromUser) {
        // TODO Auto-generated method stub
        tv1.setText("当前值为:"+progress);
      }
      public void onStartTrackingTouch(SeekBar seekBar) {
        // TODO Auto-generated method stub
        tv2.setText("音量调节");
      }
      public void onStopTrackingTouch(SeekBar seekBar) {
        // TODO Auto-generated method stub
        tv2.setText("音量调节结束");
      }
    }
}
```

图 3-21 拖动条的使用

3.2.4 界面布局

1. 普通样式

Android 常用五种布局方式，分别是：FrameLayout（框架布局），LinearLayout（线性布局），AbsoluteLayout（绝对布局），RelativeLayout（相对布局），TableLayout（表格布局）。

（1）FrameLayout：框架布局，所有内容依次都放在左上角，会重叠，这个布局比较简单，也只能放一点比较简单内容。

（2）LinearLayout：线性布局，每一个 LinearLayout 里面又可分为垂直布局（android:orientation="vertical"）和水平布局（android:orientation="horizontal"）。垂直布局时，每一行就只有一个元素，多个元素依次垂直往下；水平布局时，只有一行，每一个元素依次向右排列。

应用举例 Example_3_21——垂直线性布局。布局为 LinearLayout，方向为 android:orientation="vertical"，添加 4 个 textview，设置文字大小和背景颜色，如图 3-22 所示。

应用举例 Example_3_22——水平线性布局。布局为 LinearLayout，方向为 android:orientation

="horizontal",添加 4 个 textview,设置文字大小和背景颜色,如图 3-23 所示。

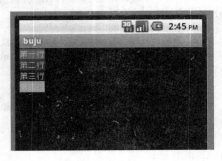

图 3-22 垂直线性布局

图 3-23 水平线性布局

(3)AbsoluteLayout:绝对布局用 X,Y 坐标来指定元素的位置,这种布局方式也比较简单,但是在屏幕旋转时,往往会出问题,而且多个元素的时候,计算比较麻烦。布局中各种控件可以随意摆放位置,这种布局方式也比较简单,但是在屏幕旋转时,往往会出问题,而且多个元素的时候,计算比较麻烦。

(4)RelativeLayout:相对布局可以理解为以某一个元素为参照物,来定位的布局方式。相对布局可以在一行或一列上显示多个控件,相对布局允许子元素指定他们相对于其他元素或父元素的位置(通过 ID 指定)。

应用举例 Example_3_23——布局为 RelativeLayout 添加 1 个 TextView,1 个 EditText,2 个 Button,方位为 TextView 在最左上方,EditText 在 TextView 下面,"fill parent",两个 Button 都在 EditText 下方,Button2(取消)靠右,Button1(确定)在 Button2(取消)的左边,如图 3-24 所示。

(5)TableLayout:表格布局,每一个 TableLayout 里面有表格行 TableRow,TableRow 里面可以具体定义每一个元素。每一个布局都有自己适合的方式,这五个布局元素可以相互嵌套应用,做出美观的界面。

应用举例 Example_3_24——布局 TableLayout,设定 android:stretchColumns="1"(1 列),在 TableLayout 中添加 6 个 TableRow(6 行),每行中添加 2 个 TextView,如图 3-25 所示。

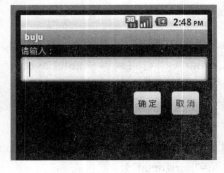

图 3-24 相对布局

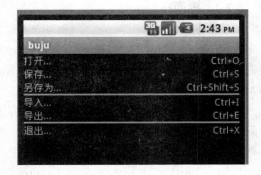

图 3-25 表格布局

 提示

"分割线"用 View 表示:
`<View`

```
android:layout_height="2dip"
android:background="#ff909090">
</View>
```

2. 高级样式

（1）切换卡（Tabwidget）：

TabWidget 类似于 Android 中查看电话簿的界面，通过多个标签切换显示不同的内容，如图 3-26 所示。

TabHost 是一个用来存放多个 Tab 标签的容器，每个 Tab 都可以对应自己的布局。

图 3-26 Android 中查看电话簿的界面

TabHost 有两个 children，一组使用户可以选择指定 Tab 页的标签，另一组是 FrameLayout 用来显示 Tab 页的内容。

1）TabHost 重要方法：

addTab( )：添加一项 Tab 页。

clearAllTabs( )：清除所有与之相关联的 Tab 页。

getCurrentTab( )：返回当前 Tab 页。

getTabContentView( )：返回包含内容的 FrameLayout。

newTabSpec( )：返回一个与之关联的新的 TabSpec。

2）Tab 重要方法：

setIndicator：设置 Tab 名称、样式。

setContent：关联控件。

（2）切换卡的使用方法：

通过 getTabHost 方法来获取 TabHost 的对象。

通过 addTab 方法来向 TabHost 中添加 Tab。

切换 Tab 时产生一个事件，要捕捉这个事件需要设置 TabActivity 的事件监听 setOnTabChangedListener。

注意直接追加 TABHOST 布局，其中增加 xmlns:android="http://schemas.android.com/apk/res/android" 即可。

应用举例 Example_3_25——切换卡的使用，如图 3-27 所示。

图 3-27 切换卡的使用

3.3 任 务 实 施

3.3.1 任务一：七彩文字

1. 任务描述

每按一次"点击变色"按钮，文字变一次颜色，蓝青灰绿红黄黑，实现七彩文字效果。

2. 界面设计（见图 3-28）

图 3-28 七彩文字界面

3. 代码实现

```java
public class MainActivity extends Activity {
  TextView tv_SevenColor;//声明文本框
  Button btn_Change;//声明按钮
  int i=0;
   public void onCreate(Bundle savedInstanceState) {
      super.onCreate(savedInstanceState);
      setContentView(R.layout.activity_main);
      tv_SevenColor = (TextView)findViewById(R.id.textView2); //连接控件
      btn_Change = (Button)findViewById(R.id.button1);
      btn_Change.setOnClickListener(new changeColorListen()); //设置监听
   }
   class changeColorListen implements OnClickListener
                                        //创建监听事件,改变字体颜色
   {
      public void onClick(View v) {
        switch (i) {
        case 0:
          //设置字体颜色
          tv_SevenColor.setTextColor(Color.BLUE);  //蓝色
          break;
        case 1:
```

```
            tv_SevenColor.setTextColor(Color.CYAN);      //青色
            break;
        case 2:
            tv_SevenColor.setTextColor(Color.GRAY);      //灰色
            break;
        case 3:
            tv_SevenColor.setTextColor(Color.GREEN);     //绿色
            break;
        case 4:
            tv_SevenColor.setTextColor(Color.RED);       //红色
            break;
        case 5:
            tv_SevenColor.setTextColor(Color.YELLOW);    //黄色
            break;
        case 6:
            tv_SevenColor.setTextColor(Color.BLACK);     //黑色
            break;
        }
        if(i < 7)
        {
            i++;
        }
        else
        {
            i = 0;
        }
    }
}
```

3.3.2 任务二：红心 A 猜测

1. 任务描述

添加三个图片视图，三个文本框，一个按钮。单击一张图片，如果是红心 A，在上面显示"猜对了"，如果不是，在上面显示"猜错了"，并在下面显示已玩次数和胜率，在最下面有一个"再玩一次"按钮。

图 3-29 红心 A 猜测界面

2. 界面设计（见图 3-29）

3. 代码实现

```
public class MainActivity extends Activity {
    ImageView imageView1,imageView2,imageView3;       //声明变量
    TextView textView1,textView2;
    Button button;
    int[] image={
        R.drawable.p01,R.drawable.p02,R.drawable.p03
    };
    int x=0,y=0,sum;//x:猜对次数  y:猜错次数  sum:游戏总次数
    public void onCreate(Bundle savedInstanceState) {
        super.onCreate(savedInstanceState);
        setContentView(R.layout.activity_main);
        textView1 = (TextView)findViewById(R.id.textView1); //提示对错文本框
```

```java
        textView2 = (TextView)findViewById(R.id.textView2); //提示场次及胜率
        textView2.setText("共进行游戏:"+x+"次,胜率为:0");
        imageView1 = (ImageView)findViewById(R.id.imageView1);
        imageView2 = (ImageView)findViewById(R.id.imageView2);
        imageView3 = (ImageView)findViewById(R.id.imageView3);
        button = (Button)findViewById(R.id.button1);                //重玩按钮
        xiPai();
        imageView1.setOnClickListener(new imageView1Listener());
        imageView2.setOnClickListener(new imageView2Listener());
        imageView3.setOnClickListener(new imageView3Listener());
        button.setOnClickListener(new replayListener());//对重玩按钮设置监听
    }
    /**
     * 洗牌功能
     * 第一次循环,把第一张图片与随机一张交换
     * 第二次循环,把第二张图片与随机一张交换
     * 第三次循环,把第三张图片与随机一张交换
     * 经过三次循环,三次交换完成洗牌
     * */
    public void xiPai()
    {
        for(int i = 0; i < 3; i++)
        {
            int tmp = image[i];
            int s = (int)(Math.random()*3);         //调用随机函数  取值范围:0~1之
                                                    // 间乘3取整数部分
            image[i] = image[s];
            image[s] = tmp;
        }
    }
    class imageView1Listener implements OnClickListener
    {
      @SuppressWarnings("deprecation")
      public void onClick(View v) {
        //设置要显示的图片
        imageView1.setImageResource(image[0]);
        imageView2.setImageResource(image[1]);
        imageView3.setImageResource(image[2]);
        //设置透明度
        imageView1.setAlpha(255);
        imageView2.setAlpha(100);
        imageView3.setAlpha(100);
        //设置ImageView不可选
        imageView1.setEnabled(false);
        imageView2.setEnabled(false);
        imageView3.setEnabled(false);
        if(image[0] == R.drawable.p01)
        {
            textView1.setText("你猜对了^o^");
            x++;
        }
```

```java
            else
            {
                textView1.setText("你猜错了 o_O!");
                y++;
            }
        }
    }
    class imageView2Listener implements OnClickListener
    {
        @SuppressWarnings("deprecation")
        public void onClick(View v) {
            imageView1.setImageResource(image[0]);
            imageView2.setImageResource(image[1]);
            imageView3.setImageResource(image[2]);
            imageView1.setAlpha(100);
            imageView2.setAlpha(255);
            imageView3.setAlpha(100);
            imageView1.setEnabled(false);
            imageView2.setEnabled(false);
            imageView3.setEnabled(false);
            if(image[1] == R.drawable.p01)
            {
                textView1.setText("你猜对了^o^");
                x++;
            }
            else
            {
                textView1.setText("你猜错了 o_O!");
                y++;
            }
        }
    }
    class imageView3Listener implements OnClickListener
    {
        @SuppressWarnings("deprecation")
        public void onClick(View v) {
            // TODO Auto-generated method stub
            //设置要显示的图片
            imageView1.setImageResource(image[0]);
            imageView2.setImageResource(image[1]);
            imageView3.setImageResource(image[2]);
            imageView1.setAlpha(100);
            imageView2.setAlpha(100);
            imageView3.setAlpha(255);
            imageView1.setEnabled(false);
            imageView2.setEnabled(false);
            imageView3.setEnabled(false);
            if(image[2] == R.drawable.p01)
            {
                textView1.setText("你猜对了^o^");
                x++;
```

```
            }
            else
            {
                textView1.setText("你猜错了 o_O!");
                y++;
            }
        }
    }
    //重玩按钮的监听事件
    class replayListener implements OnClickListener
    {
        @SuppressWarnings("deprecation")
        public void onClick(View v) {
            // TODO Auto-generated method stub
            //设置要显示的图片
            imageView1.setImageResource(R.drawable.p04);
            imageView2.setImageResource(R.drawable.p04);
            imageView3.setImageResource(R.drawable.p04);
            //对图片视图设置透明度
            imageView1.setAlpha(255);
            imageView2.setAlpha(255);
            imageView3.setAlpha(255);
            //设置图片视图可选
            imageView1.setEnabled(true);
            imageView2.setEnabled(true);
            imageView3.setEnabled(true);
            textView1.setText("猜猜哪个是红心 A");
            //重新洗牌
            xiPai();
            sum = x + y;
            textView2.setText("共进行游戏:" + sum + "次,胜率为:" + x * 1.0 / sum);
        }
    }
```

3.3.3 任务三：电子相册

1. 任务描述

实现一个电子相册，拖曳鼠标可以看到下一张图片，单击鼠标会弹出图片名称。

2. 界面设计（见图 3-30）

3. 代码实现

（1）Dzxc.java 文件。

```
public class dzxc extends Activity {
    /** Called when the activity is first created. */
    Gallery ga;
    int [] aa=new int[]
                {
        R.drawable.p1,R.drawable.p2,R.drawable.p3,R.drawable.p4,R.drawable.p5
                };
    String [] bb=new String[]
```

图 3-30 电子相册界面

```java
                        {
        "动漫1","动漫2","动漫3","动漫4","动漫5"
                        };
    @Override
    public void onCreate(Bundle savedInstanceState) {
        super.onCreate(savedInstanceState);
        setContentView(R.layout.main);
         //初始化控件
        ga=(Gallery)findViewById(R.id.grid1);
        //设置适配器
        img adp=new img(dzxc.this,aa);
        ga.setAdapter(adp);
        //监听Gallery选择和滚动事件
        ga.setOnItemClickListener(new galleryListener());
    }
    class galleryListener implements OnItemClickListener
    {
       public void onItemClick(AdapterView<?> arg0, View arg1, int arg2,
            long arg3) {
        // TODO Auto-generated method stub
        //ga.setBackgroundResource(aa[arg2]);
         //通过toast显示信息
        Toast.makeText(dzxc.this, bb[arg2], Toast.LENGTH_SHORT).show();
       }
    }
}
```

（2）Img.java。

```java
public class img extends BaseAdapter {
   Context mc;
   int[] mid;
   public img(Context c,int[] id)
   {
      mc=c;
      mid=id;
   }
   public int getCount() {
      //  返回绑定列长度
      return mid.length;
   }
   public Object getItem(int position) {
      // 返回绑定类信息
      return position;
   }
   public long getItemId(int position) {
      // 返回绑定列位置
      return position;
   }
   public View getView(int position, View convertView, ViewGroup parent) {
      // 返回绑定列样式
      ImageView iv=new ImageView(mc);
```

```
            // 图片大小
    iv.setLayoutParams(new Gallery.LayoutParams(240,240));
    // 设置图片透明度
    iv.setAlpha(180);
        //设置图片位置
    iv.setScaleType(ImageView.ScaleType.FIT_CENTER);
        //设置图片资源
    iv.setImageResource(mid[position]);
    return iv;
    }
}
```

3.3.4 任务四：游戏登录界面

1. 任务描述

制作一个游戏的菜单（包含游戏登录界面），运行程序，先进入一个用户登录对话框，输入正确的用户名和密码允许登录，然后进入主菜单界面，主菜单界面包括"开始游戏"和"读取进度"两项内容，单击"开始游戏"，进入一个环形进度条对话框，显示"游戏初始化中"，15 秒后关闭，单击"读取进度"，进入一个柱形进度条对话框控制柱形的状态，每 500 毫秒改变 2%，显示"游戏进度读取中"。

2. 界面设计（见图 3-31）

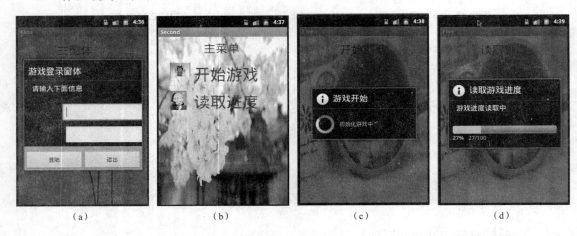

图 3-31 游戏登录界面
（a）登录窗体；（b）主菜单界面；（c）游戏初始化界面；（d）进度读取界面

3. 代码实现

（1）base.java 文件。

```
public class base extends BaseAdapter {
    Context mc;
    int ma[];
    String mb[];
    base(Context c,int a[],String b[])
    {
        mc=c;
        ma=a;
```

```java
        mb=b;
    }
    public int getCount() {
        // 返回信息长度
        return ma.length;
    }
    public Object getItem(int position) {
        // 返回信息内容
        return ma[position];
    }
    public long getItemId(int position) {
        // 返回信息的位置
        return position;
    }
    public View getView(int position, View convertView, ViewGroup parent) {
        //返回信息的样式
        //当前是界面时可以get布局,当前不是界面只能from应用布局
        LayoutInflater li=LayoutInflater.from(mc);
        View v=li.inflate(R.layout.zjm_bj, null);
        ImageView iv=(ImageView)v.findViewById(R.id.imageView1);
        TextView tv=(TextView)v.findViewById(R.id.textView1);
        iv.setImageResource(ma[position]);
        tv.setText(mb[position]);
        return v;
    }
}
```

（2）First.java 文件。

```java
public class First extends Activity {
    Builder ag;
    View vv;
    public void onCreate(Bundle savedInstanceState) {
        super.onCreate(savedInstanceState);
        setContentView(R.layout.activity_first);
        cjct();
    }
    public void cjct()
    {
        //初始化游戏登录窗体
        LayoutInflater li=getLayoutInflater();
        vv=li.inflate(R.layout.zct_ct, null);
        ag=new AlertDialog.Builder(First.this);
        ag.setTitle("游戏登录窗体");
        ag.setMessage("请输入下面信息");
        ag.setView(vv);
        ag.setPositiveButton("登录", new jt1());
        ag.setNegativeButton("退出", new jt2());
        ag.create();
        ag.show();
    }
    class jt1 implements OnClickListener
```

```java
        {
          public void onClick(DialogInterface dialog, int which) {
            // 游戏登录判定
            EditText et1,et2;
            et1=(EditText)vv.findViewById(R.id.edit1);
            et2=(EditText)vv.findViewById(R.id.edit2);
            if(et1.getText().toString().equals("admin")&&et2.getText().toString().equals("654321"))
            {
              Intent t =new Intent();
              t.setClass(First.this, Second.class);
              startActivity(t);
            }
            else
            {
              Toast.makeText(First.this, "输入信息有误", Toast.LENGTH_SHORT).show();
              cjct();
            }
          }
        }
        class jt2 implements OnClickListener
        {
          public void onClick(DialogInterface dialog, int which) {
            // TODO Auto-generated method stub
            First.this.finish();
          }
        }
        @Override
        public boolean onCreateOptionsMenu(Menu menu) {
            getMenuInflater().inflate(R.menu.activity_first, menu);
            return true;
        }
    }
```

（3）Second.java 文件。

```java
public class Second extends Activity {
    ListView lv;
    int img[]={
        R.drawable.tb1,R.drawable.tb2
    };
    String mm[]={
        "开始游戏","读取进度"
    };
    @Override
    public void onCreate(Bundle savedInstanceState) {
        super.onCreate(savedInstanceState);
        setContentView(R.layout.activity_second);
        lv=(ListView)findViewById(R.id.listView1);
        base bs=new base(Second.this, img, mm);
        lv.setAdapter(bs);
        lv.setOnItemClickListener(new jt());
```

```java
    }
    class jt implements OnItemClickListener
    {
      public void onItemClick(AdapterView<?> arg0, View arg1, int arg2,
          long arg3) {
        // TODO Auto-generated method stub
        Intent t=new Intent();
        t.setClass(Second.this, Five.class);
        t.putExtra("bz", arg2);
        startActivity(t);
      }
    }
    @Override
    public boolean onCreateOptionsMenu(Menu menu) {
        getMenuInflater().inflate(R.menu.activity_second, menu);
        return true;
    }
    @Override
    public boolean onMenuItemSelected(int featureId, MenuItem item) {
        // 设置主菜单实现跳转界面
        Intent t=new Intent();
        switch(item.getItemId())
        {
          case R.id.item1:
              t.setClass(Second.this, Four.class);
              startActivity(t);
              break;
          case R.id.item2:
              t.setClass(Second.this, Third.class);
              startActivity(t);
              break;
        }
        return true;
    }
}
```

（4）Third.java 文件。

```java
public class Third extends Activity {

    @Override
    public void onCreate(Bundle savedInstanceState) {
        super.onCreate(savedInstanceState);
        setContentView(R.layout.activity_third);
    }
    @Override
    public boolean onCreateOptionsMenu(Menu menu) {
        getMenuInflater().inflate(R.menu.activity_third, menu);
        return true;
    }
}
```

（5）Four.java 文件。

```java
public class Four extends Activity {

    @Override
    public void onCreate(Bundle savedInstanceState) {
        super.onCreate(savedInstanceState);
        setContentView(R.layout.activity_four);
    }
    @Override
    public boolean onCreateOptionsMenu(Menu menu) {
        getMenuInflater().inflate(R.menu.activity_four, menu);
        return true;
    }
}
```

（6）Five.java 文件。

```java
public class Five extends Activity {
    TextView tv;
    ProgressDialog pd;
    int m=0;
    public void onCreate(Bundle savedInstanceState) {
        super.onCreate(savedInstanceState);
        setContentView(R.layout.activity_five);
        //当从主窗体跳转到游戏界面时,传递掉转方式值bz
        tv=(TextView)findViewById(R.id.textView1);
        Intent t=getIntent();
        int n=t.getIntExtra("bz", 2);
        //表示游戏开始
        if(n==0)
        {
           tv.setText("开始游戏");
           pd=new ProgressDialog(Five.this);
           pd.setTitle("游戏开始");
           pd.setMessage("初始化游戏中");
           pd.setProgressStyle(ProgressDialog.STYLE_SPINNER);
           pd.show();
           new Thread()
           {
              public void run()
              {
                try {
                  sleep(10000);
                } catch (InterruptedException e) {
                  // TODO Auto-generated catch block
                  e.printStackTrace();
                }
                pd.dismiss();
              }
           }.start();
```

```java
//读取进度
if(n==1)
{
   tv.setText("读取进度");
   pd=new ProgressDialog(Five.this);
   pd.setTitle("读取游戏进度");
   pd.setMessage("游戏进度读取中");
   pd.setProgressStyle(ProgressDialog.STYLE_HORIZONTAL);
   pd.setProgress(100);
   pd.show();

   new Thread()
   {
      public void run()
      {
         try {
            while(m<100)
            {
             sleep(100);
             pd.setProgress(m++);
            }
         } catch (InterruptedException e) {
            // TODO Auto-generated catch block
            e.printStackTrace();
         }
           pd.dismiss();
      }
   }.start();
  }
}
@Override
public boolean onCreateOptionsMenu(Menu menu) {
    getMenuInflater().inflate(R.menu.activity_five, menu);
    return true;
}
}
```

第 4 章 Android 手机功能开发

▲ 引言

Android 平台主要应用于移动智能终端平台，并为手机端进行特别的优化，开发者在学习 Android 开发时，需要着重学习 Android 在移动智能终端的特性。本章主要介绍 Android 手机功能开发方面的知识，通过介绍 Android 四大组件中 Intent 组件的基本使用，Android 手机中短信、电话等功能的实现方法，通过广播和资源管理器的学习掌握 Android 环境下手机资源的使用方式，通过实训项目来实现对手机功能开发的整体掌握。

4.1 Intent 组件的使用

Intent 是一个将要执行的动作的抽象的描述，一般来说是作为参数来使用，由 Intent 来协助完成 Android 各个组件之间的通讯。

4.1.1 Intent 的组成

1. 动作（action）

用来指明要实施的动作是什么，比如说 ACTION_VIEW, ACTION_EDIT 等，常用动作说明如表 4-1 所示。

表 4-1 常 用 动 作 说 明 表

常量名称	常 量 值	意 义
ACTION_MAIN	android.intent.action.MAIN	应用程序入口
ACTION_VIEW	android.intent.action.VIEW	显示数据给用户
ACTION_ATTACH_DATA	android.intent.action.ATTACH_DATA	指明附加信息给其他地方的一些数据
ACTION_EDIT	android.intent.action.EDIT	显示可编辑的数据
ACTION_PICK	android.intent.action.PICK	选择数据
ACTION_CHOOSER	android.intent.action.CHOOSER	显示一个 Activity 选择器
ACTION_GET_CONTENT	android.intent.action.GET_CONTENT	获得内容
ACTION_DIAL	android.intent.action.GET_CONTENT	显示打电话面板
ACITON_CALL	android.intent.action.DIAL	直接打电话
ACTION_SEND	android.intent.action.SEND	直接发短信
ACTION_SENDTO	android.intent.action.SENDTO	选择发短信

续表

常量名称	常量值	意义
ACTION_ANSWER	android.intent.action.ANSWER	应答电话
ACTION_INSERT	android.intent.action.INSERT	插入数据
ACTION_DELETE	android.intent.action.DELETE	删除数据
ACTION_RUN	android.intent.action.RUN	运行数据
ACTION_SYNC	android.intent.action.SYNC	同步数据
ACTION_PICK_ACTIVITY	android.intent.action.PICK_ACTIVITY	选择 Activity
ACTION_SEARCH	android.intent.action.SEARCH	搜索
ACTION_WEB_SEARCH	android.intent.action.WEB_SEARCH	Web 搜索
ACTION_FACTORY_TEST	android.intent.action.FACTORY_TEST	工厂测试入口点

2. 数据（data）

要事实的具体的数据，一般由一个 Uri 变量来表示。

提示

Intent 要完成的操作是由动作+数据决定的，如表 4-2 所示。

表 4-2　　　　　　　　**Intent 的 Action 和 Data 属性匹配说明表**

Action 属性	Data 属性	说明
ACTION_VIEW	content://contacts/people/1	显示 ID 为 1 的联系人信息
ACTION_DIAL	content://contacts/people/1	将 ID 为 1 的联系人电话号码显示在拨号界面中
ACITON_VIEW	tel:123	显示电话为 123 的联系人信息
ACITON_VIEW	http://www.google.com	在浏览器中浏览该网站
ACITON_VIEW	file://sdcard/mymusic.mp3	播放 MP3
ACITON_VIEW	geo:39.2456，116.3523	显示地图

3. 分类（Category）

这个选项指定了将要执行的这个 action 的其他一些额外的信息，例如 LAUNCHER_CATEGORY，表示 Intent 的接受者应该在 Launcher 中作为顶级应用出现；而 ALTERNATIVE_CATEGORY 表示当前的 Intent 是一系列的可选动作中的一个，这些动作可以在同一块数据上执行。

4. 类型（Type）

显式指定 Intent 的数据类型（MIME）。一般 Intent 的数据类型能够根据数据本身进行判定，但是通过设置这个属性，可以强制采用显式指定的类型而不再进行推导。

5. 组件（Component）

指定 Intent 的目标组件的类名称。通常 Android 会根据 Intent 中包含的其他属性的信息，比如 action、data/type、category 进行查找，最终找到一个与之匹配的目标组件。但是，如果 component 这个属性有指定的话，将直接使用它指定的组件，而不再执行上述查找过程。指定了这个属性以后，Intent 的其他所有属性都是可选的。

6. 扩展信息（Extra）

扩展信息是其他所有附加信息的集合。使用 extras 可以为组件提供扩展信息，比如，如果要执行"发送电子邮件"这个动作，可以将电子邮件的标题、正文等保存在 extras 里，传给电子邮件发送组件。

4.1.2 Intent 寻找目标组件的方法

1. Intent 启动不同组件的方法

表 4-3　　　　　　　　　　　　　Intent 启动不同组件的方法

组件名称	方法名称
Activity（界面跳转）	startActivity( )
	startActivityForResult( )
Service（启动服务）	startService( )
	bindService( )
Broadcasts（进行广播）	sendBroadcast( )
	sendOrderedBroadcast( )
	sendStickyBroadcast( )

2. 利用 Intent 调用 Android 系统资源

调用 web 浏览器：

Uri uri = Uri.parse("http://www.google.com");

Intent it = new Intent(Intent.ACTION_VIEW,uri);

startActivity(it);

利用 google 搜索：

Intent intent = new Intent( );

intent.setAction(Intent.ACTION_WEB_SEARCH);

intent.putExtra(SearchManager.QUERY,"searchString")

startActivity(intent);

显示地图：

Uri uri = Uri.parse("geo:38.899533,−77.036476");

Intent it = new Intent(Intent.Action_VIEW,uri);

startActivity(it);

地图路径规划：

Uri uri = Uri.parse("http://maps.google.com/maps?" +

"f=dsaddr=startLat startLng&daddr=endLat endLng&hl=en");

Intent it = new Intent(Intent.ACTION_VIEW,URI);

startActivity(it);

拨打电话：

Uri uri = Uri.parse("tel:xxxxxx");

Intent it = new Intent(Intent.ACTION_DIAL, uri);

startActivity(it);
调用发短信的程序：
Intent it = new Intent(Intent.ACTION_VIEW);
it.putExtra("sms_body", "The SMS text");
it.setType("vnd.android-dir/mms-sms");
startActivity(it);
发送短信：
Uri uri = Uri.parse("smsto:0800000123");
Intent it = new Intent(Intent.ACTION_SENDTO, uri);
it.putExtra("sms_body", "The SMS text");
startActivity(it);
发送彩信：
Uri uri = Uri.parse("content://media/external/images/media/23");
Intent it = new Intent(Intent.ACTION_SEND);
it.putExtra("sms_body", "some text");
it.putExtra(Intent.EXTRA_STREAM, uri);
it.setType("image/png");
startActivity(it);
发送 Email：
Uri uri = Uri.parse("mailto:xxx@abc.com");
Intent it = new Intent(Intent.ACTION_SENDTO, uri);
startActivity(it);
播放多媒体：
Intent it = new Intent(Intent.ACTION_VIEW);
Uri uri = Uri.parse("file:///sdcard/song.mp3");
it.setDataAndType(uri, "audio/mp3");
startActivity(it);
卸载应用程序：
Uri uri = Uri.fromParts("package", strPackageName, null);
Intent it = new Intent(Intent.ACTION_DELETE, uri);
startActivity(it);
安装应用程序：
Uri installUri = Uri.fromParts("package", "xxx", null);
returnIt = new Intent(Intent.ACTION_PACKAGE_ADDED, installUri);
打开照相机：
Intent i = new Intent(Intent.ACTION_CAMERA_BUTTON, null);
this.sendBroadcast(i);
从 gallery 选取图片：
Intent i = new Intent( );

```
i.setType("image/*");
i.setAction(Intent.ACTION_GET_CONTENT);
startActivityForResult(i, 11);
```
打开录音机：
```
Intent mi = new Intent(Media.RECORD_SOUND_ACTION);
    startActivity(mi);
```
打开另一程序：
```
Intent i = new Intent();
ComponentName cn = new ComponentName("com.yellowbook.android2",
        "com.yellowbook.android2.AndroidSearch");
i.setComponent(cn);
i.setAction("android.intent.action.MAIN");
startActivityForResult(i, RESULT_OK);
```

应用举例 Example_4_1——创建一个 Android 应用程序，实现手机基本功能，实现手机上网、发短信、打电话、发送邮件功能，具体步骤如下：

（1）完成程序的界面设计：界面上添加 4 个 EditText 控件分别用于输入网址、电话号码、短信内容、邮箱地址。添加 4 个 Button 控件实现打开网页、发短信、拨打电话、发送邮件的事件触发，如图 4-1 所示。

（2）在 src 目录下完成应用程序的事件处理，本程序源代码如下：

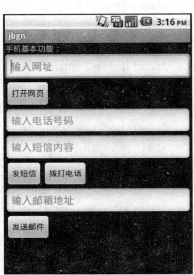

图 4-1 Example_4_1 界面效果

```java
import android.app.Activity;
import android.content.Intent;
import android.net.Uri;
import android.os.Bundle;
import android.view.View;
import android.view.View.OnClickListener;
import android.widget.Button;
import android.widget.EditText;
public class jbgn extends Activity {
    /** Called when the activity is first created. */
  Button bt1,bt2,bt3,bt4;
  EditText et1,et2,et3,et4;
   @Override
   public void onCreate(Bundle savedInstanceState) {
       super.onCreate(savedInstanceState);
       setContentView(R.layout.main);
       //初始化相关控件
       bt1=(Button)findViewById(R.id.button1);
       bt2=(Button)findViewById(R.id.button2);
       bt3=(Button)findViewById(R.id.button3);
       bt4=(Button)findViewById(R.id.button4);
       et1=(EditText)findViewById(R.id.editText1);
```

```java
            et1.setHint("输入网址");
            et2=(EditText)findViewById(R.id.editText2);
            et2.setHint("输入电话号码");
            et3=(EditText)findViewById(R.id.editText3);
            et3.setHint("输入短信内容");
            et4=(EditText)findViewById(R.id.editText4);
            et4.setHint("输入邮箱地址");
            //设置监听器
            bt1.setOnClickListener(new jt1());
            bt2.setOnClickListener(new jt2());
            bt3.setOnClickListener(new jt3());
            bt4.setOnClickListener(new jt4());
    }
    class jt1 implements OnClickListener
    {
        public void onClick(View v) {
            //打开网页
            Uri uri = Uri.parse("http://"+et1.getText().toString());
            Intent it = new Intent(Intent.ACTION_VIEW,uri);
            startActivity(it);
        }
    }
    class jt2 implements OnClickListener
    {
        public void onClick(View v) {
            //拨打电话
            Uri uri = Uri.parse("tel:"+et2.getText().toString());
            Intent it = new Intent(Intent.ACTION_DIAL, uri);
            startActivity(it);
        }
    }
    class jt3 implements OnClickListener
    {
        public void onClick(View v) {
            //发送短信
            Uri uri = Uri.parse("smsto:"+et2.getText().toString());
            Intent it = new Intent(Intent.ACTION_SENDTO, uri);
            it.putExtra("sms_body", et3.getText().toString());
            startActivity(it);
        }
    }
    class jt4 implements OnClickListener
    {
        public void onClick(View v) {
          //发电子邮件
            Uri uri = Uri.parse("mailto:"+et4.getText().toString());
            Intent it = new Intent(Intent.ACTION_SENDTO, uri);
            startActivity(it);
        }
    }
}
```

4.2 Android 广播机制

4.2.1 Android 的广播机制介绍

当 Android 系统发生一些特定的事件，这些事件需要被其他应用程序获知时，android 系统会通过广播的方式发送这个事件，如图 4-2 所示。

系统发出广播后，应用程序会通过自己编写的广播接收器来判断是否要接收这些事件，当需要接收时则进行相应的事件处理。

通过广播机制，程序员通过相关广播事件，决定要完成的任务，比如设计手机电话黑名单软件时，可以监听系统电话广播事件，当发现系统电话广播时，与预设

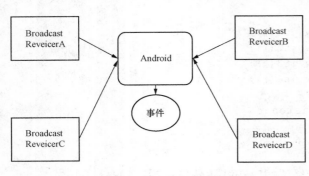

图 4-2　Android 广播机制

的黑名单进行比较，匹配时挂掉用户的电话；设计大头贴功能时，可以监听系统的手机拍摄事件，当发现手机拍摄事件时可调转软件对拍摄照片进行处理。

4.2.2 广播机制的实现方法

根据电台的广播原理相同，Android 的广播机制，可以理解为电台在某个频率发送广播信号，用户则接收固定频率的广播信号，接收和发送广播可以在不同的应用程序中，发送的广播事件可以是自定义的广播事件也可以是系统的广播事件，同样接收广播的事件可以接收自定义的广播事件也可以接收系统的广播事件。

编写广播事件：

（1）创建 Itent 对象；

（2）通过 setAction 设置广播类型（系统或自定义广播）；

（3）通过 sendBroadcast 放送广播。

编写接收广播：

（1）新建一个类，继承 Broadcastreceiver；

（2）复写 onReceive 方法（要实现的功能）；

（3）在 manifest 进行注册（使用 receiver 标签注册，intent-filter 起到过滤器功能，通过 action 来判断接收哪种广播事件）。

应用举例 Example_4_2——创建一个 Android 应用程序，实现发送自定义广播、系统广播，实现对不同广播的接收，具体步骤如下：

（1）完成程序的界面设计：界面上添加两个 Button 控件实现发送系统广播和自定义广播，如图 4-3 所示。

（2）src 目录下完成应用程序的事件处理，本程序源代码如下：

图 4-3　Example_4_2 界面效果

界面对应的类文件中，完成不同广播事件的发送源代码如下：

```java
import android.app.Activity;
import android.content.Intent;
import android.os.Bundle;
import android.view.View;
import android.view.View.OnClickListener;
import android.widget.Button;
public class gb extends Activity {
   Button bt1,bt2;
    public void onCreate(Bundle savedInstanceState) {
        super.onCreate(savedInstanceState);
        setContentView(R.layout.main);
        bt1=(Button)findViewById(R.id.button1);
        bt2=(Button)findViewById(R.id.button2);
        bt1.setOnClickListener(new jt1());
        bt2.setOnClickListener(new jt2());
    }
    class jt1 implements OnClickListener
    {
      public void onClick(View v) {
        //发送系统原有的动作：具有一定通用性,有可能会触发某些特殊的时间
        Intent t=new Intent();
        //发送的动作为：显示可编辑的数据
        t.setAction(Intent.ACTION_EDIT);
        gb.this.sendBroadcast(t);
      }
    }
    class jt2 implements OnClickListener
    {
      public void onClick(View v) {
        //发送自定义的动作：通用性比较差,特定群接受信息,防止系统的错误
        Intent t=new Intent();
        t.setAction("wodedongzuo");
        gb.this.sendBroadcast(t);
      }
    }
}
```

创建一个新的类，接收系统广播，源代码如下：

```java
import android.content.BroadcastReceiver;
import android.content.Context;
import android.content.Intent;
import android.widget.Toast;
public class jsxtxx extends BroadcastReceiver{
   public void onReceive(Context context, Intent intent) {
      //接受系统信息的接收器
      Toast.makeText(context, "接受到了Action_edit信息", Toast.LENGTH_LONG).show();
   }
}
```

创建一个新的类，接收自定义广播，源代码如下：

```
import android.content.BroadcastReceiver;
import android.content.Context;
import android.content.Intent;
import android.widget.Toast;
public class jszdyxx extends BroadcastReceiver {
    public void onReceive(Context context, Intent intent) {
        // 接收自定义的广播接收器
        Toast.makeText(context, "接收到了自定义信息", Toast.LENGTH_LONG).show();
    }
}
```

(3) 需要在 Androidmanifest.xml 文件中，对应的 application 标签声明中追加代码：

```
<application android:icon="@drawable/icon" android:label="@string/app_name">
    <activity android:name=".gb"
              android:label="@string/app_name">
        <intent-filter>
            <action android:name="android.intent.action.MAIN" />
            <category android:name="android.intent.category.LAUNCHER" />
        </intent-filter>
    </activity>
    <receiver android:name=".jsxtxx">
        <intent-filter>
            <action android:name="android.intent.action.EDIT" />
        </intent-filter>
    </receiver>
    <receiver android:name=".jszdyxx">
        <intent-filter>
            <action android:name="wodedongzuo"/>
        </intent-filter>
    </receiver>
</application>
```

提 示

需要进行广播接收时必须进行接收注册，接收器注册的方式有两种：可以在 AndroidManifest.xml 当中进行注册，或者直接在程序代码中进行注册。

4.3 Android 资源管理器

在开发应用程序时经常需要利用一些系统提供的资源，Android 对于这些系统资源采用资源管理器方式进行管理。

活动管理器： 管理应用程序的系统状态

权限：`<uses-permission android:name="android.permission.GET_TASKS"/>`

代码：`ActivityManager activityManager = (ActivityManager) getSystemService(Context.ACTIVITY_SERVICE);`

警报管理器：闹钟的服务

权限：

代码：AlarmManager alarmManager = (AlarmManager) getSystemService(Context.ALARM_SERVICE);

音频管理器：

权限：

代码：AudioManager audioManager = (AudioManager) getSystemService(Context.AUDIO_SERVICE);

剪贴板管理器：

权限：

代码：ClipboardManager clipboardManager = (ClipboardManager) getSystemService(Context.CLIPBOARD_SERVICE);

连接管理器：网络连接的服务

权限：<uses-permission android:name="android.permission.ACCESS_NETWORK_STATE"/>

代码：ConnectivityManager connectivityManager = (ConnectivityManager) getSystemService(Context.CONNECTIVITY_SERVICE);

输入法管理器：

权限：

代码：InputMethodManager inputMethodManager = (InputMethodManager) getSystemService(Context.INPUT_METHOD_SERVICE);

键盘管理器：键盘锁的服务

权限：

代码：KeyguardManager keyguardManager = (KeyguardManager) getSystemService(Context.KEYGUARD_SERVICE);

布局解压器管理器取得 xml 里定义的 view

权限：

代码：LayoutInflater layoutInflater = (LayoutInflater) getSystemService(Context.LAYOUT_INFLATER_SERVICE);

位置管理器：位置的服务，如 GPS

权限：

代码：LocationManager locationManager = (LocationManager) getSystemService(Context.LOCATION_SERVICE);

通知管理器：状态栏的服务

权限：

代码：NotificationManager notificationManager = (NotificationManager) getSystemService(Context.NOTIFICATION_SERVICE);

电源管理器：电源的服务

权限：<uses-permission android:name="android.permission.DEVICE_POWER"/>

代码：PowerManager powerManager = (PowerManager) getSystemService(Context.POWER_SERVICE);

搜索管理器：搜索的服务

权限：

代码：SearchManager searchManager = (SearchManager) getSystemService(Context.SEARCH_SERVICE);

传感器管理器

权限：

代码：SensorManager sensorManager = (SensorManager) getSystemService(Context.SENSOR_SERVICE);

电话管理器：电话服务

权限：<uses-permission android:name="android.permission.READ_PHONE_STATE"/>

代码：TelephonyManager telephonyManager = (TelephonyManager) getSystemService(Context.TELEPHONY_SERVICE);

振动器：手机振动的服务

权限：<uses-permission android:name="android.permission.VIBRATE"/>

代码：Vibrator vibrator = (Vibrator) getSystemService(Context.VIBRATOR_SERVICE);

墙纸：

权限：<uses-permission android:name="android.permission.SET_WALLPAPER"/>

代码：WallpaperService wallpaperService = (WallpaperService) getSystemService(Context.WALLPAPER_SERVICE);

Wi-Fi 管理器：Wi-Fi 服务

权限：<uses-permission android:name="android.permission.ACCESS_WIFI_STATE"/>

代码：WifiManager wifiManager = (WifiManager) getSystemService(Context.WIFI_SERVICE);

窗口管理器：管理打开的窗口程序

权限：

代码：WindowManager windowManager = (WindowManager) getSystemService(Context.WINDOW_SERVICE);

4.3.1 音频管理器的使用

AudioManager：音频管理器，控制和访问铃声（调节声音大小、静音、振动）。

（1）AudioManager 常用方法。

adjustVolume：调整音量大小。

getMode：得到当前音频模式。

getRingerMode：得到当前铃声模式。

getStreamMaxVolume：得到最大的音频指数。

getStreamVolume：得到当前音频指数。

isSpeakerphoneOn：扬声器是否打开。

loadSoundEffects：加载声音效果。

playSoundEffect：播放声音效果。

setMicrophoneMute：设置麦克风静音和关闭。

setMode：设置音频模式。

setRingerMode：设置铃声模式。

setSpeakerphoneOn：设置扬声器打开和关闭。

setVibrateSetting：当铃声模式改变时，设置振动类型。

（2）AudioManager 常用信息。

ADJUST_LOWER：降低铃声音量。

ADJUST_RAISE：增加铃声音量。

RINGER_MODE_NORMAL：响铃模式。

STREAM_RING：铃声流（通过它获得当前音量）。

RINGER_MODE_SILENT：静音模式。

RINGER_MODE_VIBRATE：振动模式。

手机铃声的设置：

振动：setRingerMode(AudioManager.RINGER_MODE_VIBRATE);

静音：setRingerMode(AudioManager.RINGER_MODE_SILENT);

铃声：setRingerMode(AudioManager.RINGER_MODE_NORMAL);

提高音量：adjustVolume(AudioManager.ADJUST_RAISE, 0);

降低音量：adjustVolume(AudioManager.ADJUST_LOWER, 0);

获得当前手机音量：getStreamVolume(AudioManager. STREAM_RING)。

应用举例 Example_4_3——创建一个 Android 应用程序，使用 progressbar 显示音量，使用 5 个按钮进行不同的音量设置（增大、减少、静音、振动、响铃），具体步骤如下。

（1）完成程序的界面设计：界面上添加 5 个 Button 控件实现音量的增大、减少、静音、振动、响铃功能，添加 progressbar 实现显示当前音量功能，如图 4-4 所示。

图 4-4　Example_4_3 界面效果

（2）src 目录下完成应用程序的事件处理，本程序源代码如下：

```
import android.media.AudioManager;
import android.os.Bundle;
import android.app.Activity;
import android.content.Context;
import android.view.Menu;
import android.view.View;
import android.view.View.OnClickListener;
import android.widget.Button;
import android.widget.ProgressBar;
public class MainActivity extends Activity {
    Button bt1,bt2,bt3,bt4,bt5;
    AudioManager am;
    ProgressBar pb;
    public void onCreate(Bundle savedInstanceState) {
        super.onCreate(savedInstanceState);
        setContentView(R.layout.activity_main);
        bt1=(Button)findViewById(R.id.button1);
        bt2=(Button)findViewById(R.id.button2);
```

```java
        bt3=(Button)findViewById(R.id.button3);
        bt4=(Button)findViewById(R.id.button4);
        bt5=(Button)findViewById(R.id.button5);
        pb=(ProgressBar)findViewById(R.id.progressBar1);
        am=(AudioManager) getSystemService(Context.AUDIO_SERVICE);
        bt1.setOnClickListener(new jt1());
        bt2.setOnClickListener(new jt2());
        bt3.setOnClickListener(new jt3());
        bt4.setOnClickListener(new jt4());
        bt5.setOnClickListener(new jt5());
        pb.setMax(7);
        pb.setProgress(am.getStreamVolume(AudioManager.STREAM_RING));
    }
    class jt1 implements OnClickListener
    {
      public void onClick(View v) {
          // 设置为静音
          am.setRingerMode(AudioManager.RINGER_MODE_SILENT);
      }
    }
    class jt2 implements OnClickListener
    {
      public void onClick(View v) {
          // 设置为振动
          am.setRingerMode(AudioManager.RINGER_MODE_VIBRATE);
          am.setVibrateSetting(AudioManager.VIBRATE_TYPE_RINGER,  AudioManager.VIBRATE_SETTING_ON);
      }
    }
    class jt3 implements OnClickListener
    {
      public void onClick(View v) {
          // 设置为响铃
          am.setRingerMode(AudioManager.RINGER_MODE_NORMAL);
      }
    }
    class jt4 implements OnClickListener
    {
      public void onClick(View v) {
          //增大音量
          am.adjustVolume(AudioManager.ADJUST_RAISE, 7);
          pb.setProgress(am.getStreamVolume(AudioManager.STREAM_RING));
      }
    }
    class jt5 implements OnClickListener
    {
      public void onClick(View v) {
          // 降低音量
          am.adjustVolume(AudioManager.ADJUST_LOWER, 7);
          pb.setProgress(am.getStreamVolume(AudioManager.STREAM_RING));
      }
```

                }
        }

4.3.2 警报管理器的使用

AlarmManager：警报管理器，专门用来设定在某个指定的时间去完成指定的事件。提供了系统警报的服务，只要在程序中设置了警报服务，通过广播接收器就可以进行接收，就算系统处于待机状态，同样不会影响运行。

取消 AlarmManager 服务：Cancel(PendingIntent operation)

设置 AlarmManager 服务：Set(int type, long triggerAtTime, PendingIntent operation)

设置周期性 AlarmManager 服务：setRepeating(int type, long triggerAtTime, long interval, PendingIntent operation)，其中：

type：RTC_WAKEUP 设置服务在系统休眠时同样会运行。

triggerAtTime：设置定时时间换算成的毫秒数。

Interval：等待时间，单位是毫秒。

PendingIntent：用来描述 intent 和这个 intent 的接收器。可以看作交给别的程序，别的程序根据这个描述在后面的时间做你安排做的事情（在合适的时候触发事件）。可以通过如下方法创建：getActivity(Context context, int requestCode, Intent intent, int flags) getBroadcast(Context context, int requestCode, Intent intent, int flags) getService(Context context, int requestCode, Intent intent, int flags)

Context：应该开始的 activity。

requestCode：传递的私有信息（一般不使用），默认 0。

Intent：意图。

Flags：传递时要做的操作，默认 0。

应用举例 Example_4_4——创建一个 Android 应用程序，通过 timepicker 设置闹钟，到达定时使用 toast 进行通知，设置周期性闹钟，每隔 20 秒 toast 提示一次，具体步骤如下：

（1）完成程序的界面设计：界面上添加 3 个 Button 控件实现设置闹钟、取消闹钟、设置周期性闹钟功能，添加 timeipcker 实现设置时间功能，如图 4-5 所示。

（2）src 目录下完成应用程序的事件处理，本程序源代码如下。

界面对应的类文件中，完成不同闹钟事件设置源代码如下：

图 4-5　Example_4_4 界面效果

```
import java.util.Calendar;
import android.app.Activity;
import android.app.AlarmManager;
import android.app.PendingIntent;
import android.content.Intent;
import android.os.Bundle;
import android.view.View;
import android.view.View.OnClickListener;
import android.widget.Button;
```

```java
import android.widget.TextView;
import android.widget.TimePicker;
import android.widget.TimePicker.OnTimeChangedListener;
public class nz extends Activity {
    TimePicker tp;
    Button bt1,bt2,bt3;
    TextView tv;
    Calendar c;
    public void onCreate(Bundle savedInstanceState) {
        super.onCreate(savedInstanceState);
        setContentView(R.layout.main);
        //初始化控件
        tp=(TimePicker)findViewById(R.id.timePicker1);
        bt1=(Button)findViewById(R.id.button1);
        bt2=(Button)findViewById(R.id.button2);
        bt3=(Button)findViewById(R.id.button3);
        tv=(TextView)findViewById(R.id.text);
        tp.setIs24HourView(true);
        c=Calendar.getInstance();
        tv.setText("请设置闹钟");
        tp.setOnTimeChangedListener(new jt());
        bt1.setOnClickListener(new jt1());
        bt2.setOnClickListener(new jt2());
        bt3.setOnClickListener(new jt3());
    }
    class jt implements OnTimeChangedListener
    {
        public void onTimeChanged(TimePicker view, int hourOfDay, int minute) {
            // 设置时间
            c.set(Calendar.HOUR_OF_DAY,hourOfDay);
            c.set(Calendar.MINUTE,minute);
            c.set(Calendar.SECOND, 0);
            c.set(Calendar.MILLISECOND,0);
        }
    }
    class jt1 implements OnClickListener
    {
        public void onClick(View v) {
            // 设置闹钟
            Intent t=new Intent();
            t.setAction("nz");
            PendingIntent pi=PendingIntent.getBroadcast(nz.this, 0, t, 0);
            AlarmManager am;
            am=(AlarmManager) getSystemService(ALARM_SERVICE);
            am.set(AlarmManager.RTC_WAKEUP, c.getTimeInMillis(), pi);
            tv.setText("定时器的时间为:"+c.get(Calendar.HOUR_OF_DAY)+":"+c.get(Calendar.MINUTE)+":"+c.get(Calendar.SECOND));
        }
    }
    class jt2 implements OnClickListener
    {
```

```java
        public void onClick(View v) {
            //取消闹钟
            Intent t=new Intent();
            t.setAction("nz");
            PendingIntent pi=PendingIntent.getBroadcast(nz.this, 0, t, 0);
            AlarmManager am;
            am=(AlarmManager) getSystemService(ALARM_SERVICE);
            am.cancel(pi);
             tv.setText("请设置闹钟");
        }
    }
    class jt3 implements OnClickListener
    {
        public void onClick(View v) {
            //设置周期性闹钟
            Intent t=new Intent();
            t.setAction("nz");
            PendingIntent pi=PendingIntent.getBroadcast(nz.this, 0, t, 0);
            AlarmManager am;
            am=(AlarmManager) getSystemService(ALARM_SERVICE);
            am.setRepeating(AlarmManager.RTC_WAKEUP, c.getTimeInMillis(), 10000, pi);
            tv.setText(" 周期性定时器的时间为:"+c.get(Calendar.HOUR_OF_DAY)+":"+c.get(Calendar.MINUTE)+":"+c.get(Calendar.SECOND));
        }
    }
}
```

创建类，设置接收闹钟事件源代码如下：

```java
import android.content.BroadcastReceiver;
import android.content.Context;
import android.content.Intent;
import android.widget.Toast;
public class jsq extends BroadcastReceiver{
   public void onReceive(Context context, Intent intent) {
     //当闹钟触发时,执行的操作
      Toast.makeText(context, "时间到了", Toast.LENGTH_LONG).show();
   }
}
```

（3）需要在 Androidmanifest.xml 文件中，对应的 application 标签声明中追加代码：

```xml
<application android:icon="@drawable/icon" android:label="@string/app_name">
    <activity android:name=".nz"
              android:label="@string/app_name">
       <intent-filter>
          <action android:name="android.intent.action.MAIN" />
          <category android:name="android.intent.category.LAUNCHER" />
       </intent-filter>
    </activity>
    <receiver android:name=".jsq">
        <intent-filter>
           <action android:name="nz" />
```

```
            </intent-filter>
        </receiver>
</application>
```

4.4 手机基本功能开发

4.4.1 短信功能

完成短信功能开发时，需要拥有相应权限才能使用。

图 4-6 Example_4_5 界面效果

发送短信：
权限：android.permission.SEND_SMS
SmsManager：短信管理器
初始化：SmsManager.getDefault( );
发送短信：sendTextMessage (电话, null, 内容, null, null);
应用举例 Example_4_5——创建一个 Android 应用程序，实现发送短信功能，具体步骤如下。

（1）完成程序的界面设计：界面上添加两个 EditText 控件存放电话和短信内容，添加 1 个 Button 控件实现发送短信功能，如图 4-6 所示。

（2）src 目录下完成应用程序的事件处理，本程序源代码如下：

```
import android.os.Bundle;
import android.app.Activity;
import android.telephony.SmsManager;
import android.view.Menu;
import android.view.View;
import android.view.View.OnClickListener;
import android.widget.Button;
import android.widget.EditText;

public class MainActivity extends Activity {
    EditText et1,et2;
    Button bt;
    public void onCreate(Bundle savedInstanceState) {
        super.onCreate(savedInstanceState);
        setContentView(R.layout.activity_main);
        et1=(EditText)findViewById(R.id.editText1);
        et2=(EditText)findViewById(R.id.editText2);
        bt=(Button)findViewById(R.id.button1);
        bt.setOnClickListener(new jt());
    }
    class jt implements OnClickListener
    {
      public void onClick(View v) {
        // 发送短信功能
        SmsManager sm=SmsManager.getDefault();
```

第4章 Android 手机功能开发

```
        sm.sendTextMessage(et1.getText().toString(), null, et2.getText().
toString(), null, null);
        }
    }
}
```

(3) 需要在 Androidmanifest.xml 文件中，追加权限：

```
<uses-permission android:name="android.permission.SEND_SMS" />
```

> **提 示**
>
> 开发电话功能软件时，可通过打开多个模拟器进行测试，模拟器显示的编号就是对应的电话号码。

接收短信：

权限：android.permission.RECEIVE_SMS

接收器（动作）：android.provider.Telephony.SMS_RECEIVED

应用举例 Example_4_6——创建一个 Android 应用程序，实现接收短信功能，通过 Toast 显示发送短信的号码和短信内容，具体步骤如下。

（1）src 目录下完成应用程序的事件处理，建立短信接收器，本程序源代码如下：

```
import android.content.BroadcastReceiver;
import android.content.Context;
import android.content.Intent;
import android.os.Bundle;
import android.telephony.gsm.SmsManager;
import android.telephony.gsm.SmsMessage;
import android.widget.Toast;
public class js extends BroadcastReceiver{
    public void onReceive(Context context, Intent intent) {
        //接收短信,解析短信的内容,用toast展示出来
        //1) 获取intent中所有数据,bundle 存储
        Bundle b=intent.getExtras();
        //2) 提取bundle 中pdus 格式短信信息,非一条短信
        Object[] p=(Object[])b.get("pdus");
        //3) 获取我们需要的内容:smsmessage
        SmsMessage[] sms=new SmsMessage[p.length];
        //4) 用smsmessage 接收短信
        for(int i=0;i<p.length;i++)
        {
            //从pdus 中解析内容赋给smsmeaage
            sms[i]=SmsMessage.createFromPdu((byte[])p[i]);
        }
        //需要当前短信中,发送端电话,发送来的短信内容
        StringBuilder s1=new StringBuilder();
        StringBuilder s2=new StringBuilder();
        for(int i=0;i<sms.length;i++)
        {
            //获取发送端电话号
            s1.append(sms[i].getOriginatingAddress());
            //收到的短信内容
```

```
            s2.append(sms[i].getMessageBody());
        }
        Toast.makeText(context, s1.toString()+" 说 :"+s2.toString(), Toast.
LENGTH_LONG).show();
    }
}
```

(2) 需要在 Androidmanifest.xml 文件中，追加权限，并注册短信接收器：

```
<?xml version="1.0" encoding="utf-8"?>
<manifest xmlns:android="http://schemas.android.com/apk/res/android"
    package="com.example.al1"
    android:versionCode="1"
    android:versionName="1.0" >
    <uses-sdk
        android:minSdkVersion="10"
        android:targetSdkVersion="10" />
    <uses-permission android:name="android.permission.RECEIVE_SMS"/>
    <application
        android:allowBackup="true"
        android:icon="@drawable/ic_launcher"
        android:label="@string/app_name"
        android:theme="@style/AppTheme" >
        <activity
            android:name="com.example.al1.MainActivity"
            android:label="@string/app_name" >
            <intent-filter>
                <action android:name="android.intent.action.MAIN" />
                <category android:name="android.intent.category.LAUNCHER" />
            </intent-filter>
        </activity>
        <receiver android:name=".js">
            <intent-filter>
                <action android:name="android.provider.Telephony.SMS_RECEIVED"/>
            </intent-filter>
        </receiver>
    </application>
</manifest>
```

4.4.2　电话功能

1. 拨打电话功能

权限：android.permission.CALL_PHONE

提　示

开发手机电话功能时 Android 并未给用户提供相关的开发接口，一般采用 intent 调用系统界面完成，如果需要独立完成电话功能开发，需要调用第三方控件完成相关开发。

应用举例 Example_4_7——创建一个 Android 应用程序，实现拨打电话功能，具体步骤如下。

（1）完成程序的界面设计：界面上添加 1 个 EditText 控件存放电话内容，添加 1 个 Button 控件实现拨打电话功能，如图 4-7 所示。

（2）src 目录下完成应用程序的事件处理，本程序源代码如下：

```
import android.net.Uri;
import android.os.Bundle;
import android.app.Activity;
import android.content.Intent;
import android.view.Menu;
import android.view.View;
import android.view.View.OnClickListener;
import android.widget.Button;
import android.widget.EditText;
public class MainActivity extends Activity {
    EditText et;
    Button bt;
    public void onCreate(Bundle savedInstanceState) {
        super.onCreate(savedInstanceState);
        setContentView(R.layout.activity_main);
        et=(EditText)findViewById(R.id.editText1);
        bt=(Button)findViewById(R.id.button1);
        bt.setOnClickListener(new jt());
    }
    class jt implements OnClickListener
    {
      public void onClick(View v) {
        // 拨打电话功能
        Uri u=Uri.parse("tel:"+et.getText().toString());
        Intent t=new Intent(Intent.ACTION_CALL, u);
        t.setFlags(Intent.FLAG_ACTIVITY_NEW_TASK);
        startActivity(t);
      }
    }
}
```

图 4-7 Example_4_7 界面效果

（3）需要在 Androidmanifest.xml 文件中，追加权限：

`<uses-permission android:name="android.permission.CALL_PHONE"/>`

2. 接收电话消息

权限：android.permission.READ_PHONE_STATE

接收器（动作）：android.intent.action.PHONE_STATE

应用举例 Example_4_7——创建一个 Android 应用程序，实现拨打电话功能，具体步骤如下。

（1）src 目录下完成应用程序的事件处理，建立短信接收器，本程序源代码如下：

```
import android.content.BroadcastReceiver;
import android.content.Context;
import android.content.Intent;
import android.widget.Toast;
public class jsq extends BroadcastReceiver {
    public void onReceive(Context context, Intent intent) {
```

```
        // TODO Auto-generated method stub
        Toast.makeText(context, "你收到电话信息", Toast.LENGTH_LONG).show();
    }
}
```

（2）需要在 Androidmanifest.xml 文件中，追加权限，并注册短信接收器：

```xml
<manifest xmlns:android="http://schemas.android.com/apk/res/android"
    package="com.example.receivephone"
    android:versionCode="1"
    android:versionName="1.0">
    <uses-sdk android:minSdkVersion="10" android:targetSdkVersion="15" />
    <uses-permission android:name="android.permission.READ_PHONE_STATE"/>
    <application android:label="@string/app_name"
        android:icon="@drawable/ic_launcher"
        android:theme="@style/AppTheme">
         <receiver android:name=".jsq">
        <intent-filter>
            <action android:name="android.intent.action.PHONE_STATE"/>
        </intent-filter>
        </receiver>
    </application>
</manifest>
```

项目实训 1：手机情景模式

1. 实训目的

（1）掌握 Android 应用程序的基本开发步骤；
（2）熟悉 Eclipse 的基本操作界面，快捷键使用；
（3）掌握 Android 广播机制的实现方式；
（4）掌握 Android 资源管理器的应用方式；
（5）掌握 Android 高级布局的使用，切换卡的使用。

2. 常见问题分析

（1）实现手机情景模式的界面样式时，一般采用切换卡实现，但也可以采用其他的布局样式；
（2）实现定时修改手机情景模式功能时，需要采用警报管理器来实现；
（3）实现对手机音频调节时，需要使用音频管理器来实现；

3. 实训内容

编写 Android 下的应用程序，实现手机情景模式，要求如下。
（1）基本模式：（使用 audiomanager 可以实现对手机状态的修改）。
1）户外模式：声音、加振动。
2）室内模式：声音设置适中。
3）会议模式：手机处于振动状态。
4）静音模式：手机处于静音状态。
（2）自定义模式（使用 audiomanager 可以实现对手机状态的修改）。
设置手机状态（振动、静音、响铃）同时设置铃声音量。
（3）定时修改模式（广播：到达某个时间进行修改）。

设置一个具体的时间,到达这个时间后修改手机的状态。(例如,用户设置睡觉的时间,在这个时间点修改手机状态为静音模式)

(4)完成以上基本功能。

(5)扩展性要求:自行研究手机情景模式的其他应用功能。

参考效果如图 4-8 所示。

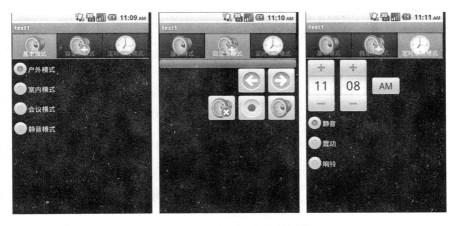

图 4-8 项目实训 1 完成效果图

4. 程序完成步骤设计

(1)界面设计:

布局:tabhost。

在不同标签下进行页面的设计。

(2)程序代码的编写:

设置 TABhost 布局。

绑定相关控件。

设置相应的监听器。

实现各个监听器的功能。

设计其他界面或者接收器。

(3)Manifest 中注册相关的组件。

(4)测试:

功能性测试。

友好性测试。

项目实训 2:短信自动回复

1. 实训目的

(1)掌握 Android 下手机基本功能的开发;

(2)熟悉 Eclipse 的基本操作界面,快捷键使用;

(3)掌握 Android 广播接收器的实现方式;

(4)掌握 Android 手机应用软件的开发流程;

(5)掌握 Android 手机短信功能的开发流程。

2. 常见问题分析
(1) 手机短信功能开发，需要程序能够完成短信信息的解析。
(2) 开发短信回复功能时，应用程序需要具备短信接收和发送短信权限。
(3) 接收短消息时，需要为应用程序建立短信接收器。
3. 实训内容
短信自动回复功能要求如下：
(1) 应用程序能够自动接收短信消息；
(2) 应用程序判定短信消息，当是指定的用户时，自动回复短信。

第 5 章 Android 数据存储开发

▲ 引言

Android 应用程序开发时,当应用程序需要进行数据存储或者使用调用系统中的数据信息时,开发者就需要掌握如何进行数据的存储和获取系统数据资源的方式。本章主要介绍在 Android 平台下数据存储和数据共享的开发方式,同时学习 Android 四大组件中数据共享组件 Content Provider 的使用,通过项目手机通讯录实现整体学习 ANDROID 的数据存储开发。

5.1 需 求 分 析

利用 Android 的 Content Provider 组件开发手机通讯录,主要实现以下功能:
(1) 实现共用手机中联系人信息;
(2) 实现插入联系人、联系人电话;
(3) 实现查询已存在的联系人信息。

5.2 知 识 准 备

5.2.1 Content Provider 组件的使用

在 Android 系统中,没有一个公共的内存区域,供多个应用共享存储数据。Android 中的 Content provider 机制可支持在多个应用中存储和读取数据。

Android 提供了一些主要数据类型的 Content provider,比如音频、视频、图片和私人通讯录等。可在 android.provider 包下面找到一些 android 提供的 Content provider。可以获得这些 Content provider,查询它们包含的数据,当然前提是已获得适当的读取权限。

1. Content Provider 的基本概念
(1) Content Provider 的作用。
1)建立访问数据统一的接口:对数据进行了封装,用户不需要存储方式就能使用数据。
2)不同应用程序间数据共享:当应用程序需要暴露自己的数据给别的应用程序时采用 content provider 进行实现。
(2) Content Provider 使用表的形式来组织数据。
1)无论数据来源什么样子都采用表的形式进行存储。
2)对于每一种类型的数据都提供了独立的表。

表 5-1　　　　　　　　　　　**Content Provider 数据组织形式**

_ID	NUMBER	NUMBER_KEY	LABEL	NAME	TYPE
13	(425)555 6677	425 555 6677	Kirkland office	Bully Pulpit	TYPE_WORK
44	(212)555-1234	212 555 1234	NY apartment	Alan Vain	TYPE_HOME
45	(212)555-6657	212 555 6657	Downtown office	Alan Vain	TYPE_MOBILE
53	201.555.4433	201 555 4433	Love Nest	Rex Cars	TYPE_HOME

（3）URI。URI 是统一资源标识符，对于 android 的每一种资源都有一个标识符，当用户需要调用某种资源时就可以使用 URI 进行访问。对于每一个 ContentProvider 都拥有一个 URI，这个 URI 用来表示这个 ContentProvider 所提供的数据。Android 提供的 ContentProvider 都存放在 android.provider 包中。

URI 由三部分组成："content：//"、数据路径、标识 ID，例如：

1）content://media/internal/images（返回设备上存储的所有图片）。

2）content://contacts/people/5（返回 Id 为 5 的联系人记录）。

3）content://contacts/people/（返回所有联系人的记录）。

要表示具体的数据，一般由一个 URI 变量来表示。

2. Content Resolver 所提供的函数

Content Resolver：实现 content provider 的接口，通过 get ContentResolver( )来获得一个 Content Resolver 对象，使用 Content Resolver 提供的方法来操作需要使用的 content provider。

（1）query( )：查询。

（2）insert( )：插入。

（3）update( )：更新。

（4）delete( )：删除。

（5）getType( )：获得数据类型。

（6）onCreate( )：创建时的函数。

3. content provider 使用过程

（1）在 androidmanifest 中设定要获取数据的权限。

（2）在程序中创建 Content Resolver 对象。

（3）定义要使用的 URI。

（4）使用 Content Resolver 提供的方法完成数据的操作。

举例 1：信息插入的步骤。

1）准备接口（getcontentresolver）。

2）准备键值对（contentvalues）。

3）定位（定位新数据在 URI 中的位置）。

4）获得该行的标志位（contenturis.parseid(uri)）。

5）设置插入数据（参考位置，数据类型，数据值）。

6）插入数据。

插入联系人信息说明：

URI：android.provider.contactscontract.data.content_uri

（空值列：rawcontacts.content_uri）

位置关键字：structuresname.raw_contact_id

类型关键字：data.mimetype 类型 structturedname.content_item_type

数据列关键字：structturedname.display_name

位置关键字：phone.raw_contact_id

类型关键字：data.mimetype 类型 phone.content_item_type

数据列关键字：phone.number

举例 2：查询的基本步骤。

1）准备接口。

2）通过 URI 查询数据。

3）读取数据。

查询联系人信息说明：

联系人姓名 URI：contactscontract.contats.content_uri

关键字：contactscontract.contats._id

姓名：contactscontract.contats.DISPLAY_NAME

电话 URI：contactscontract.commondatakinds.phone.conttent_uri

关键字：contactscontract.commondatakinds.phone.contats._id

电话号：contactscontract.commondatakinds.phone.number

5.2.2 SQLite 使用

SQLite 是一个微型数据库，提供了比较完整的关系型数据库功能，其官方网站：http://www.sqlite.org。Android 系统自行定义了一套对 SQLite 访问的方法。

1. SQLite 的一般使用步骤

（1）创建一个类继承 SQLiteOpenHelper（复写构造函数、定义数据库创建时 oncreate、定义数据库版本升级 onupgrade 方法）：要定义数据库中表结构。

（2）创建数据操作类（构造函数、打开、关闭、增、删、改、查方法）。

（3）在程序中完成数据操作。

SQLiteOpenHelper 是一个助手类，用于对数据库的创建和管理。其中包含如下的方法：

1）构造函数：传入窗体、数据库名、null、版本号。

2）oncreate：数据库创建时要做的操作（生成表结构）。

3）onupgrade：数据库版本更新时操作（通过版本号确定是否更新）。

2. 定义数据相关操作

（1）数据库打开：创建助手类对象，定义 Sqlite 的操作。

1）getReadableDatabase( )：得到一个可读的数据库（查询）。

2）geWritableDatabase( )：得到一个可写的数据库（增、删、改）。

（2）数据库关闭：调用 close( )方法。

（3）数据库信息插入：

1）contentviews values=new ContentViews( )；该对象存放的就是键值对，键是列名，值是希望插入的值。

2）values.put（"列名"，值）。
3）db.insert（"表名"，null，values）；对于未插入数据使用 null（默认值）。
（4）数据库信息更新。
1）contentviews values=new ContentViews（ ）；该对象存放的就是键值对，键是列名，值是希望插入的值。
2）values.put（"列名"，值）
3）db.update（"表名"，values，"id=？"，null），参数：表名，contentviews 对象，相当于 where 字句？占位符。
（5）数据库信息删除。调用 delete 方法实现：
delete（"表名"，"id=？"，null），参数：表名、where 字句、占位符。
（6）数据库信息查询。调用 query 方法实现：调用 query 方法会返回一个游标 cursor，使用 moveToNext( )方法下移，就可以查询相关数据。

图 5-1　SQLite 操作的主界面

获得数据：public Cursor query (String table, String[] columns, String selection, String[] selectionArgs, String groupBy, String having, String orderBy)。

其中：table：表名，columns：列名，Selection：where 子句，selectionArgs：字符转义，groupBy：groupBy 子句，having：having 子句，orderBy：orderBy 子句。

应用举例 Example_5_1——利用 SQLite 进行数据库相关操作，包含信息的增加、修改、删除、查询功能。

（1）完成程序的界面设计：本程序用 ListView 控件显示信息，利用 4 个 Button 控件实现增加、修改、删除、查询功能，本程序设置的主界面效果如图 5-1 所示。

（2）本程序源代码如下：
1）dbhelp.java 文件。

```java
import android.content.Context;
import android.database.sqlite.SQLiteDatabase;
import android.database.sqlite.SQLiteOpenHelper;

public class dbhelp extends SQLiteOpenHelper{
    private static String dbname="txl.db";
    private static int dbver=1;
    private String dbcreate="CREATE TABLE tb_lxrxx (" +
        "id integer primary key autoincrement," +
        "name text," +
        "mobiletel text)";
    public dbhelp(Context context) {
        super(context, dbname, null, dbver);
        // TODO Auto-generated constructor stub
    }
    public void onCreate(SQLiteDatabase db) {
        //数据库创建时
        db.execSQL(dbcreate);
```

```
    }
    public void onUpgrade(SQLiteDatabase db, int oldVersion, int newVersion) {
        //数据库升级时
        db.execSQL("DROP TABLE IF EXISTS tb_lxrxx");
        onCreate(db);
    }
}
```

2）mydb.java 文件。

```
import android.content.ContentValues;
import android.content.Context;
import android.database.Cursor;
import android.database.sqlite.SQLiteDatabase;

public class mydb {
    private Context mc;
    private dbhelp mdbhelp;
    private SQLiteDatabase msqlite;
    public mydb(Context c)
    {
        mc=c;
    }
    public void open()
    {
        mdbhelp=new dbhelp(mc);
        msqlite=mdbhelp.getWritableDatabase();
    }
    public void close()
    {
        mdbhelp.close();
    }
    public void insertdb(String name,String tel)
    {
        ContentValues value=new ContentValues();
        value.put("name", name);
        value.put("mobiletel", tel);
        msqlite.insert("tb_lxrxx", null, value);
    }
    public Cursor selectdb()
    {
        return msqlite.query("tb_lxrxx", new String[]{"id","name","mobiletel"}, null, null, null, null, null);
    }
    public Cursor selectdbbyid(int id)
    {
        return msqlite.query("tb_lxrxx", new String[]{"id","name","mobiletel"}, "id="+id, null, null, null, null);
    }
    public void deletedb(int id)
    {
        msqlite.delete("tb_lxrxx", "id="+id, null);
```

```
        }
        public void updatedb(int id,String name,String tel)
        {
           ContentValues value=new ContentValues();
           value.put("name", name);
           value.put("mobiletel", tel);
           msqlite.update("tb_lxrxx", value, "id="+id, null);
        }
}
```

3) czjm.java 文件（主界面）。

```
import java.util.ArrayList;

import android.app.Activity;
import android.database.Cursor;
import android.os.Bundle;
import android.view.View;
import android.view.View.OnClickListener;
import android.widget.ArrayAdapter;
import android.widget.Button;
import android.widget.ListAdapter;
import android.widget.ListView;
import android.widget.SimpleCursorAdapter;

public class czjm extends Activity {
   Button bt1,bt2,bt3,bt4,bt5;
   ListView lv;
    public void onCreate(Bundle savedInstanceState) {
       super.onCreate(savedInstanceState);
       setContentView(R.layout.main);
       //初始化控件
       bt1=(Button)findViewById(R.id.button1);
       bt2=(Button)findViewById(R.id.button2);
       bt3=(Button)findViewById(R.id.button3);
       bt4=(Button)findViewById(R.id.button4);
       bt5=(Button)findViewById(R.id.button5);
       lv=(ListView)findViewById(R.id.listView1);
       bt1.setOnClickListener(new jt1());
       bt2.setOnClickListener(new jt2());
       bt3.setOnClickListener(new jt3());
       bt4.setOnClickListener(new jt4());
       bt5.setOnClickListener(new jt5());
    }
    class jt1 implements OnClickListener
    {
      public void onClick(View v) {
         //插入信息操作
         mydb my=new mydb(czjm.this);
         my.open();
         my.insertdb("hewei", "13796210718");
         my.close();
```

第 5 章 Android 数据存储开发

```java
        }
    }
    class jt2 implements OnClickListener
    {
      public void onClick(View v) {
         //删除数据
         mydb my=new mydb(czjm.this);
         my.open();
         my.deletedb(1);
         my.close();
      }
    }
    class jt3 implements OnClickListener
    {
      public void onClick(View v) {
         //更新数据操作
         mydb my=new mydb(czjm.this);
         my.open();
         my.updatedb(2, "lisi", "12345678");
         my.close();
      }
    }
    class jt4 implements OnClickListener
    {
    //执行查询操作,并进行信息绑定
      public void onClick(View v) {
         // 查询操作
         mydb my=new mydb(czjm.this);
         my.open();
         Cursor cu=my.selectdb();
         ArrayList<String> name=new ArrayList<String>();
         while(cu.moveToNext())
         {
            name.add(cu.getString(cu.getColumnIndex("name")));
         }
         my.close();
         ArrayAdapter<String> aa=new ArrayAdapter<String>(czjm.this, android.R.layout.simple_list_item_single_choice,name);
         lv.setAdapter(aa);
      }
    }
    class jt5 implements OnClickListener
    {
      public void onClick(View v) {
         //条件查询
         mydb my=new mydb(czjm.this);
         my.open();
         Cursor cu=my.selectdbbyid(2);
         ArrayList<String> name=new ArrayList<String>();
         while(cu.moveToNext())
         {
```

```
        name.add(cu.getString(cu.getColumnIndex("name")));
    }
    my.close();
    ArrayAdapter<String> aa=new ArrayAdapter<String>(czjm.this, android.R.layout.simple_list_item_single_choice,name);
    lv.setAdapter(aa);
  }
 }
}
```

5.3 任 务 实 施

5.3.1 任务描述

利用 Content Provider 实现一个简易的通讯录，能完成联系人姓名电话的插入和查询功能。

5.3.2 实现步骤

（1）界面设计。
（2）主菜单设计。
（3）在后台进行映射控件。
（4）查询联系人姓名和电话信息的查询。
（5）将查询的信息绑定到 listview 上，注采用复写 baseadapter 方式。
（6）通过主菜单实现界面跳转。
（7）在跳转到的界面，实现联系人姓名和电话信息的插入。
（8）对整个程序进行优化设计。
（9）进行整体测试。

5.3.3 界面设计

设计通讯录界面、数据查询界面，具体界面如图 5-2 所示。

图 5-2　通讯录界面

5.3.4 代码实现

（1）MainActivity.java 文件。

```java
import java.util.ArrayList;
import java.util.ResourceBundle.Control;
import android.os.Bundle;
import android.provider.ContactsContract;
import android.app.Activity;
import android.content.ContentResolver;
import android.content.Intent;
import android.database.Cursor;
import android.view.Menu;
import android.view.MenuItem;
import android.widget.ListView;
public class MainActivity extends Activity {
    ListView lv;
    ArrayList<String> name=new ArrayList<String>();
    ArrayList<String> number=new ArrayList<String>();
    protected void onCreate(Bundle savedInstanceState) {
        super.onCreate(savedInstanceState);
        setContentView(R.layout.activity_main);
        lv=(ListView) findViewById(R.id.listView1);
        sjbd();
        base bs=new base(MainActivity.this, name, number);
        lv.setAdapter(bs);
    }
    private void sjbd() {
        // 查询相关数据,将查询结果写入动态链表
        ContentResolver cr=getContentResolver();
        //获取联系人的基本信息
        Cursor cur=cr.query(ContactsContract.Contacts.CONTENT_URI,
                    null,       null,       null,           null);
        while(cur.moveToNext())
        {
            //当联系人不为空时,读取关键ID和联系人姓名
            String id=cur.getString(cur.getColumnIndex(ContactsContract.Contacts._ID));
            String lxr=cur.getString(cur.getColumnIndex(ContactsContract.Contacts.DISPLAY_NAME));
            //使用关键ID查询对应的电话信息
            Cursor pcur=cr.query(ContactsContract.CommonDataKinds.Phone.CONTENT_URI,
                    null,
                    ContactsContract.CommonDataKinds.Phone.CONTACT_ID+"="+id,
                    null,
                    null);
            String num="";
            while(pcur.moveToNext())
            {
                num=pcur.getString(pcur.getColumnIndex(ContactsContract.CommonDataKinds.Phone.NUMBER));
            }
```

```java
            //把姓名与电话写入动态链表
            System.out.println(lxr);
            name.add(lxr);
            number.add(num);
            //关闭使用的游标
            pcur.close();
        }
         cur.close();
    }
    public boolean onCreateOptionsMenu(Menu menu) {
        // Inflate the menu; this adds items to the action bar if it is present.
        getMenuInflater().inflate(R.menu.main, menu);
        return true;
    }
    public boolean onOptionsItemSelected(MenuItem item) {
        if(item.getItemId()==R.id.action_settings)
        {
            //界面跳转
            Intent t=new Intent();
            t.setClass(MainActivity.this, Add_lxr.class);
            startActivity(t);
        }
        return super.onOptionsItemSelected(item);
    }
}
```

（2）base.java 文件。

```java
import java.util.ArrayList;
import android.content.Context;
import android.view.LayoutInflater;
import android.view.View;
import android.view.ViewGroup;
import android.widget.BaseAdapter;
import android.widget.TextView;
public class base extends BaseAdapter {
    Context mc;
    ArrayList<String> mname;
    ArrayList<String> mnumber;
    public base(Context c,ArrayList<String> name,ArrayList<String> number)
    {
        //构造函数:获取传递信息
        mc=c;
        mname=name;
        mnumber=number;
    }
    public int getCount() {
        // 返回链表长度
        return mname.size();
    }
    public Object getItem(int position) {
        // 返回链表中某一项内容
```

```
            return mname.get(position);
        }
        public long getItemId(int position) {
            // 返回链表某一项的 ID
            return position;
        }
        public View getView(int position, View convertView, ViewGroup parent) {
            //返回某一项样式
            //准备适配器,在界面上准备
            LayoutInflater li=LayoutInflater.from(mc);
            //准备某一行的视图
            View v=li.inflate(R.layout.item_ys, null);
            //声明某行所包含的控件
            TextView tv1=(TextView) v.findViewById(R.id.textView1);
            TextView tv2=(TextView) v.findViewById(R.id.textView2);
            //给某行对应的控件进行赋值
            tv1.setText(mname.get(position));
            tv2.setText(mnumber.get(position));
            return v;
        }
    }
```

（3）Add_lxr.java 文件。

```
import android.net.Uri;
import android.os.Bundle;
import android.provider.ContactsContract.CommonDataKinds.Phone;
import android.provider.ContactsContract.CommonDataKinds.StructuredName;
import android.provider.ContactsContract.Contacts.Data;
import android.provider.ContactsContract.RawContacts;
import android.app.Activity;
import android.content.ContentResolver;
import android.content.ContentUris;
import android.content.ContentValues;
import android.content.Intent;
import android.view.Menu;
import android.view.View;
import android.view.View.OnClickListener;
import android.widget.Button;
import android.widget.EditText;
public class Add_lxr extends Activity {
    //向 android 中插入一个联系人信息
    EditText et1,et2;                           //联系人信息
    Button bt;                                  //单击 BUTTON 增加联系人
    protected void onCreate(Bundle savedInstanceState) {
        super.onCreate(savedInstanceState);
        setContentView(R.layout.activity_add_lxr);
        et1=(EditText) findViewById(R.id.editText1);
        et2=(EditText) findViewById(R.id.editText2);
        bt=(Button) findViewById(R.id.button1);
        bt.setOnClickListener(new jt());
    }
```

```java
class jt implements OnClickListener
{
  public void onClick(View v) {
    //增加联系人信息
    //准备操作的游标
    ContentResolver cr=getContentResolver();
    //准备插入的数据
    ContentValues cv=new ContentValues();
    //插入前的准备:数据信息插入的位置
    //确定位置:插入一条空的数据,会返回这条空数据的位置
    Uri rawContactUri=cr.insert(RawContacts.CONTENT_URI, cv);
    long id=ContentUris.parseId(rawContactUri);
    //向刚才插入的空记录中,插入名字信息
    cv.clear();
    //ID的位置
    cv.put(StructuredName.RAW_CONTACT_ID, id);
    //插入数据的类型
    cv.put(Data.MIMETYPE, StructuredName.CONTENT_ITEM_TYPE);
    //插入的数据
    cv.put(StructuredName.DISPLAY_NAME, et1.getText().toString());
    //进行数据插入操作
    cr.insert(android.provider.ContactsContract.Data.CONTENT_URI, cv);
    //插入电话信息
    cv.clear();
    //ID的位置
    cv.put(StructuredName.RAW_CONTACT_ID, id);
    //插入数据的类型
    cv.put(Data.MIMETYPE,Phone.CONTENT_ITEM_TYPE);
    //插入的数据
    cv.put(Phone.NUMBER, et2.getText().toString());
    //进行数据插入操作
    cr.insert(android.provider.ContactsContract.Data.CONTENT_URI, cv);
    //跳回主界面
    Intent t=new Intent();
    t.setClass(Add_lxr.this, MainActivity.class);
    startActivity(t);
  }
}
```

(4) AndroidManifest.xml 文件。

```xml
<manifest xmlns:android="http://schemas.android.com/apk/res/android"
    package="com.example.al_txl"
    android:versionCode="1"
    android:versionName="1.0" >
    <uses-sdk
        android:minSdkVersion="10"
        android:targetSdkVersion="10" />
    <uses-permission android:name="android.permission.WRITE_CONTACTS"/>
    <uses-permission android:name="android.permission.READ_CONTACTS"/>
    <application
```

```xml
        android:allowBackup="true"
        android:icon="@drawable/ic_launcher"
        android:label="@string/app_name"
        android:theme="@style/AppTheme" >
        <activity
            android:name="com.example.al_txl.MainActivity"
            android:label="@string/app_name" >
            <intent-filter>
                <action android:name="android.intent.action.MAIN" />
                <category android:name="android.intent.category.LAUNCHER" />
            </intent-filter>
        </activity>
        <activity
            android:name="com.example.al_txl.Add_lxr"
            android:label="@string/title_activity_add_lxr" >
        </activity>
    </application>
</manifest>
```

第 6 章　Android 文 件 管 理

▲ 引言

智能手机一个主要的功能就是能够对手机中的文件进行管理，开发者就需要掌握如何进行 Android 的文件管理。本章主要介绍在 Android 平台下文件 I/O 操作，同时掌握关于 Android 的权限管理机制的使用方式，并通过项目来分析不同知识的使用。

6.1　需 求 分 析

利用 Android 的文件管理开发文件管理器，主要实现以下功能：
（1）实现查看手机目录中的信息；
（2）能够新建文件夹；
（3）对于文件可以进行打开、重命名、复制、剪切、删除操作。

6.2　知 识 准 备

6.2.1　动态数组

静态数组，定义时已经确定了数组的元素个数，实际开发中数组长度经常不可预知，因此要采用动态数组的方式实现。

1. 动态数组的声明

在 Java 中采用 ArrayList 方式实现动态数组。

声明：ArrayList<类型> a=new ArrayList<类型>( );

举例：ArrayList<String> aa=new ArrayList<String>( )
　　　ArrayList<Integer> tp=new ArrayList<Integer>( )

2. 动态数组常用的方法

（1）add（元素值）：增加一个新的数组元素。
（2）clear( )：清空动态数组。
（3）get（位置）：通过元素位置获得数组中元素值。
（4）size( )：元素的个数。

应用举例 Example_6_1——利用动态数组实现在 textview 中追加信息的功能。
（1）界面设计（见图 6-1）：

（2）源代码：

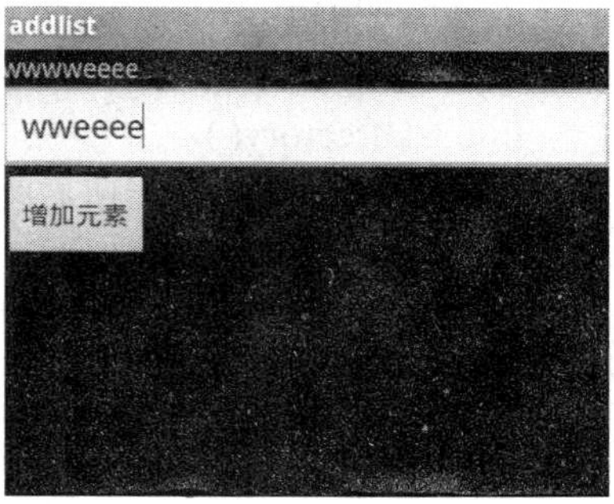

```java
import java.util.ArrayList;
import android.app.Activity;
import android.os.Bundle;
import android.view.View;
import android.view.View.OnClickListener;
import android.widget.Button;
import android.widget.EditText;
import android.widget.TextView;
public class list extends Activity {
    /** Called when the activity is first created. */
    EditText et;
    TextView tv;
    Button bt;
    ArrayList<String> name=new ArrayList<String>();
     public void onCreate(Bundle savedInstanceState) {
        super.onCreate(savedInstanceState);
        setContentView(R.layout.main);
        et=(EditText)findViewById(R.id.editText1);
        tv=(TextView)findViewById(R.id.textView1);
        bt=(Button)findViewById(R.id.button1);
        bt.setOnClickListener(new jt());
    }
    class jt implements OnClickListener
    {
      public void onClick(View v) {
        // TODO Auto-generated method stub
        name.add(et.getText().toString());
        StringBuilder sb=new StringBuilder();
        for(int i=0;i<name.size();i++)
        {
            sb=sb.append(name.get(i));
        }
        tv.setText(sb.toString());
      }
    }
}
```

图 6-1　Example_6_1 效果图

6.2.2　文件操作常用方法

1. 基本操作

（1）创建文件对象：

File file=new File("路径");

（2）获得文件的相对路径：

String path=File.getPath( );

（3）获得文件绝对路径：

String path=File.getAbsoultePath( );

（4）获得文件或文件夹的父目录：

String path=File.getParent( );

（5）获得文件或者文件夹的名字：

String name=File.getName( );

（6）判断当前是文件还是文件夹：

File.isDirectory( )

（7）列出文件夹下面所有文件和文件夹名：

File[] files=File.listFiles( );

如何实现不同类型文件显示不同的图标？

String.endsWith(part1)：字符串结尾与 part1 是否匹配。

2. 对文件进行修改的操作

注意：权限问题，对文件进行操作时需要引入下面的权限。

```
<uses-permission android:name="android.permission.WRITE_EXTERNAL_STORAGE">
</uses-permission>
```

（1）重命名：

File.renameTo（路径）

（2）文件删除：

File.delete( )

（3）新建文件：

File.createNewFile( )

（4）新建文件夹：

File.mkDir( )

（5）剪切功能：实质就是重命名。

（6）文件的打开：进行文件打开操作时，通常是通过 Intent 调用系统已有的功能界面来实现。

1）Intent t = new Intent( );

2）t.setAction(android.content.Intent.ACTION_VIEW);

3）File file = new File(aFile.getAbsolutePath( ));

4）String fileName = file.getName( );

5）根据文件结尾扩展名判定文件类型。

6）t.setDataAndType(Uri.fromFile(file), "audio/*");设置数据和文件类型

7）startActivity(t);

注意：图片："image/*"、音频："audio/*"、视频："video/*"

（7）文件复制功能：采用 Java 带缓存的输入/输出流方式实现：

1）`InputStream in=new FileInputStream(src)`//把源文件信息写入到输入信息流中。

2）`OutputStream out=new FileOutputStream(target)`//把目标文件作为输出信息流的接收方。

3）`BufferedInputStream bin=new BufferedInputStream(in)`//设置输入流的缓存区。

4）`BufferedOutputStream bout=new BufferedOutputStream(out)`//设置输出流的缓存区。

5）`byte[] b = new byte[8192]` //对文件进行分块时每块文件的大小

6）`int len = bin.read(b);`//根据块的大小确定当前文件有多少块信息。

7）`while (len != -1){`//判定是否还有文件块未被读取。

8）`bout.write(b, 0, len);`//把输入缓存区中最前面块信息写入到输出缓存区中。

9）`len = bin.read(b);`// 根据块的大小确定当前文件有多少块信息。

```
}
bin.close();//关闭输入流。
bout.close();//关闭输出流。
```

6.3 任 务 实 施

6.3.1 任务描述

利用 Android 的文件操作实现一个手机文件夹管理器，查看文件，并对文件进行相关操作：打开、删除、重命名、复制、剪切等操作。

6.3.2 实现步骤

（1）界面设计。
（2）主菜单设计。
（3）在后台进行映射控件。
（4）获取当前目录的文件列表信息。
（5）将获取的文件列表信息绑定到 listview 上，注采用复写 baseadapter 方式。
（6）监听 listview 的单击处理事件，确定文件与文件夹的不同操作。
（7）完成对文件的相关操作：打开、删除、重命名、复制、剪切等操作。
（8）完成主菜单设置的相关操作：创建文件夹，粘贴操作。
（9）进行应用程序的测试。
（10）进行应用程序优化。

图 6-2 所示为文件管理器操作基本流程图。

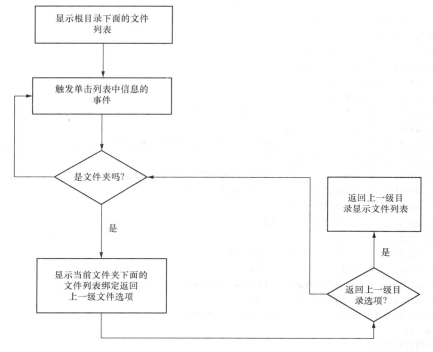

图 6-2　文件管理器操作基本流程图

6.3.3 界面设计

设计程序主界面、文件操作界面,具体界面如图 6-3 所示:

图 6-3 文件管理器界面

6.3.4 代码实现

(1) MainActivity.java 文件。

```
import java.io.BufferedInputStream;
import java.io.BufferedOutputStream;
import java.io.File;
import java.io.FileInputStream;
import java.io.FileNotFoundException;
import java.io.FileOutputStream;
import java.io.IOException;
import java.io.InputStream;
import java.io.OutputStream;
import java.util.ArrayList;
import android.net.Uri;
import android.os.Bundle;
import android.app.Activity;
import android.app.AlertDialog;
import android.app.AlertDialog.Builder;
import android.content.DialogInterface;
import android.content.DialogInterface.OnClickListener;
import android.content.Intent;
import android.speech.tts.TextToSpeech.OnInitListener;
import android.view.LayoutInflater;
import android.view.Menu;
import android.view.MenuItem;
import android.view.View;
import android.widget.AdapterView;
import android.widget.EditText;
import android.widget.AdapterView.OnItemClickListener;
import android.widget.ListView;
public class MainActivity extends Activity {
    View v;
    //声明控件的容器
    ListView lv;
```

```java
//某行要绑定文件名字,文件类型(图片)
ArrayList<String> fname=new ArrayList<String>();      //文件
ArrayList<Integer> tname=new ArrayList<Integer>();   //文件类型,用图片表示
//创建一个文件操作的数组
String fopr[]={"打开","删除","重命名","复制","剪切"};
//初始化当前文件夹位置
File nowfloder=new File("/");
//初始化当前文件的位置
File nowfile;
//要剪切或者赋值的文件
File tempfile;
int zt;//1 表示剪切,2 表示赋值
protected void onCreate(Bundle savedInstanceState) {
    super.onCreate(savedInstanceState);
    setContentView(R.layout.activity_main);
    lv=(ListView) findViewById(R.id.listView1);
    //方法:获取当前目录下的文件列表
    broseto(nowfloder);
    //单击 lv 的监听
    lv.setOnItemClickListener(new jt1());
}
class jt1 implements OnItemClickListener
{
    //当单击某一项的操作
    public void onItemClick(AdapterView<?> arg0, View arg1, int arg2,
        long arg3) {
        // TODO Auto-generated method stub
        if(tname.get(arg2)==R.drawable.up)
        {
            nowfloder=new File(nowfloder.getParent());
            broseto(nowfloder);
        }
        else
        {
            if(tname.get(arg2)==R.drawable.wjj)
            {
                if(nowfloder.getParent()==null)
                {
                    //根目录时做的文件路径转换
                    nowfloder=new File(nowfloder.getAbsolutePath()+fname.get(arg2));
                    broseto(nowfloder);
                }
                else
                {
                    //非根目录时文件路径的转换
                    nowfloder=new File(nowfloder.getAbsolutePath()+"/"+fname.get(arg2));
                    broseto(nowfloder);
                }
            }
            else
```

```
            {
                if(nowfloder.getParent()==null)
                {
                    nowfile=new File(nowfloder.getAbsolutePath()+fname.get(arg2));
                }
                else
                {
                    nowfile=new File(nowfloder.getAbsolutePath()+"/"+fname.get(arg2));
                }
                Builder bb=new AlertDialog.Builder(MainActivity.this);
                bb.setTitle("对文件进行下列操作");
                bb.setItems(fopr, new jt2());
                bb.create();
                bb.show();
            }
        }
    }
    class jt2 implements OnClickListener
    {
        public void onClick(DialogInterface dialog, int which) {
            // TODO Auto-generated method stub
            switch(which)
            {
            case 0:openfile(nowfile);break;
            case 1:deletefile(nowfile);break;
            case 2:renamefile(nowfile);break;
            case 3:copyfile(nowfile);break;
            case 4:cutfile(nowfile);break;
            }
        }
        private void cutfile(File nowfile) {
            //剪切文件操作
            zt=1;
            tempfile=nowfile;
        }
        private void copyfile(File nowfile) {
            //复制文件操作
            zt=2;
            tempfile=nowfile;
        }
        private void renamefile(File nowfile) {
            //重命名文件操作
            //1)弹出对话框,输入要修改的名字
            //创建布局适配器
            LayoutInflater li=LayoutInflater.from(MainActivity.this);
            //准备布局样式
            v=li.inflate(R.layout.windows_ys, null);
            //准备窗体
            Builder b=new AlertDialog.Builder(MainActivity.this);
```

```java
            b.setTitle("请输入新的文件名");
            b.setView(v);
            b.setPositiveButton("确认", new win_jt1());
            b.create();
            b.show();
        }
        class win_jt1 implements OnClickListener
        {
            public void onClick(DialogInterface dialog, int which) {
                // 文件重命名操作
                EditText et=(EditText) v.findViewById(R.id.editText1);
                File newfile;
                //设置新文件的路径
                newfile=new File(nowfile.getParent()+"/"+et.getText());
                //用新文件名修改当前的文件名
                nowfile.renameTo(newfile);
                broseto(nowfloder);
            }
        }
        private void deletefile(File nowfile) {
            //删除文件操作
            //存储当前文件
            File sc=nowfile;
            //获取当前文件的父路径,做刷新准备
            nowfile=new File(nowfile.getParent());
            //删除文件
            sc.delete();
            //刷新文件夹
            broseto(nowfloder);
        }
        private void openfile(File nowfile) {
            // 打开文件操作:intent
            Intent t=new Intent();
            //查看的操作
            t.setAction(Intent.ACTION_VIEW);
            //判断文件的类型,根据文件类型,选择打开文件的方式
            if(nowfile.getName().endsWith(".MP3")||nowfile.getName().endsWith(".mp3"))
            {
                t.setDataAndType(Uri.fromFile(nowfile), "audio/*");
            }
    if(nowfile.getName().endsWith(".bmp")||nowfile.getName().endsWith(".jpg"))
            {
                t.setDataAndType(Uri.fromFile(nowfile), "image/*");
            }
            startActivity(t);
        }
    }
        private void broseto(File nowf) {
            // 在 nowf 下的文件列表:数据绑定
            //将标题设置为当前目录的路径
```

```java
        this.setTitle(nowf.getAbsolutePath());
        //获取当前文件下的文件列表
        File flist[]=nowf.listFiles();
        //将文件数组中所有文件绑定到动态数组中
        fname.clear();
        tname.clear();
        //绑定返回上一层目录
        if(nowf.getParent()!=null)
        {
            //非根目录时绑定的信息
            fname.add("返回上一层目录");
            tname.add(R.drawable.up);
        }
        //循环绑定所有文件
        for(int i=0;i<flist.length;i++)
        {
            //动态数组 fname 中增加文件名信息
            fname.add(flist[i].getName());
            //判定当前文件是文件夹还是文件
            if(flist[i].isDirectory())
            {
                //是文件夹则绑定文件夹图片
                tname.add(R.drawable.wjj);
            }
            else
            {
                //否则绑定文件图片
                tname.add(R.drawable.wj);
            }
        }
        //选取适配器,绑定动态数组
        base bs=new base(MainActivity.this, fname, tname);
        lv.setAdapter(bs);
    }
    public boolean onCreateOptionsMenu(Menu menu) {
        // Inflate the menu; this adds items to the action bar if it is present.
        getMenuInflater().inflate(R.menu.main, menu);
        return true;
    }
    public boolean onOptionsItemSelected(MenuItem item) {
        // TODO Auto-generated method stub
        int i=item.getItemId();
        switch(i)
        {
        case R.id.add_wjj:
            //创建文件夹
            //创建布局适配器
            LayoutInflater li=LayoutInflater.from(MainActivity.this);
            //准备布局样式
            v=li.inflate(R.layout.windows_ys, null);
            //准备窗体
```

```java
            Builder b=new AlertDialog.Builder(MainActivity.this);
            b.setTitle("请输入新的文件夹名");
            b.setView(v);
            b.setPositiveButton("确认", new win_jt2());
            b.create();
            b.show();
            break;
        case R.id.file_zt:
            //要放置文件的位置
            File fsrc=new File(nowfloder.getAbsolutePath()+"/"+tempfile.getName());
            if(zt==1)
            {
                //文件剪切操作

                //修改文件路径
                tempfile.renameTo(fsrc);
                broseto(nowfloder);
            }
            if(zt==2)
            {
                //文件的复制操作,数据流
                InputStream in=null;
                OutputStream out=null;
                BufferedInputStream bin=null;
                BufferedOutputStream bout=null;
                //读取要复制的文件内容
                try {
                    in=new FileInputStream(tempfile);
                    out=new FileOutputStream(fsrc);
                    //读取缓存区(I/O 优化,为复制文件增加内存缓存)
                    bin=new BufferedInputStream(in);
                    //写入缓存区
                    bout=new BufferedOutputStream(out);
                    //分步读取缓存区中的信息
                    byte bb[]=new byte[8192];
                    int len=bin.read(bb);
                    while(len!=-1)
                    {
                        bout.write(bb,0,len);
                        len=bin.read();
                    }
                } catch (FileNotFoundException e) {
                    // TODO Auto-generated catch block
                    e.printStackTrace();
                }
                //向新的文件写入信息
                    catch (IOException e) {
                    // TODO Auto-generated catch block
                    e.printStackTrace();
                }
                finally
```

```java
        {
            if(bin!=null)
            {
                try {
                    bin.close();
                } catch (IOException e) {
                    // TODO Auto-generated catch block
                    e.printStackTrace();
                }
            }
            if(bout!=null)
            {
                try {
                    bin.close();
                } catch (IOException e) {
                    // TODO Auto-generated catch block
                    e.printStackTrace();
                }
            }
            broseto(nowfloder);

        }
        break;
    }
    return super.onOptionsItemSelected(item);
}
class win_jt2 implements OnClickListener
{

    @Override
    public void onClick(DialogInterface dialog, int which) {
        //实现创建文件夹
        EditText et=(EditText) v.findViewById(R.id.editText1);
        //文件夹路径
        File newfloder=new File(nowfile.getAbsolutePath()+"/"+et.getText());
        //创建文件夹
        newfloder.mkdir();
        broseto(nowfloder);
    }
}
```

（2）base.java 文件。

```java
import java.util.ArrayList;

import android.content.Context;
import android.view.LayoutInflater;
import android.view.View;
import android.view.ViewGroup;
import android.widget.BaseAdapter;
```

```java
import android.widget.ImageView;
import android.widget.TextView;
public class base extends BaseAdapter {
    Context mc;
    ArrayList<String> mfilename;
    ArrayList<Integer> mfileclass;
    public base(Context c,ArrayList<String> filename,ArrayList<Integer> fileclass)
    {
        //接收传递来的界面中的信息
        mc=c;
        mfilename=filename;
        mfileclass=fileclass;
    }
    public int getCount() {
        // 返回链表长度
        return mfilename.size();
    }
    public Object getItem(int position) {
        // 返回链表某一项值
        return mfilename.get(position);
    }
    public long getItemId(int position) {
        // 返回链表某一项的位置
        return position;
    }
    public View getView(int position, View convertView, ViewGroup parent) {
        // 返回链表的图形样式
        //在界面上准备布局适配器
        LayoutInflater li=LayoutInflater.from(mc);
        //在某一行布局上添加对应的样式
        View v=li.inflate(R.layout.row_ys, null);
        //在样式中对每一个控件进行赋值
        ImageView iv=(ImageView) v.findViewById(R.id.imageView1);
        TextView tv=(TextView) v.findViewById(R.id.textView1);
        iv.setImageResource(mfileclass.get(position));
        tv.setText(mfilename.get(position));
        return v;
    }
}
```

（3）AndroidManifest.xml 文件，增加权限：

```xml
<uses-permission android:name="android.permission.WRITE_EXTERNAL_STORAGE"/>
```

第 7 章　Android 多媒体开发

▲ 引言

多媒体即 Multimedia，它的含义是：在计算机系统中组合两种或两种以上媒体的一种人机交互式的信息交流方式，这些媒体包括文字、图片、照片、声音、动画和影片等。从多媒体的概念提出到今天，它的应用领域已经涉及了各个行业，尤其是网络技术和移动技术日益发达的今天，每个使用 Android 系统的人都知道 Android 系统中带有一个音乐播放器，音乐播放器可以看到当前终端里所有的音乐文件，这就是 Android 的多媒体。除此之外，Android 多媒体技术还支持视频的播放和录制，以及图片的采集（即拍照）。本章就来介绍如何在 Android 上应用多媒体，达到让应用程序美观、易用并且具有吸引力的效果。

7.1　需　求　分　析

利用 Android 多媒体技术开发 MP3 播放器，主要实现以下功能：
（1）实现 MP3 音乐播放；
（2）实现自动获取 sdcard 目录下的 MP3 信息；
（3）实现歌词同步问题；
（4）实现单曲循环、随机播放、顺序播放功能。

7.2　知　识　准　备

7.2.1　多媒体技术

1. MediaPlayer 类
（1）作用：用来播放音频、视频、流媒体，包含 Audio 和 Video 的播放功能。
（2）常用的方法：
1）MediaPlayer：构造方法。
2）create：创建一个要播放的多媒体。
3）getCurrentPosition：得到当前播放位置。
4）getDuration：得到文件的时间。
5）getVideoHeight：得到视频的高度。
6）getVideoWidth：得到视频的宽度。

7）isLooping：是否循环播放。
8）isPlaying：是否正在播放。
9）pause：暂停。
10）prepare：准备（同步）。
11）prepareAsync：准备（异步）。
12）release：释放 MediaPlayer 对象。
13）reset：重置 MediaPlayer 对象。
14）seekTo：制定播放的位置（以毫秒为单位的时间）。
15）setAudioStreamType：设置流媒体的类型。
16）setDataSource：设置多媒体数据来源。
17）setDisplay：设置用 SurfaceHolder 来显示多媒体。
18）setLooping：设置是否循环播放。
19）setOnBufferingUpdateListener：网络流媒体的缓冲监听。
20）setOnErrorListener：设置错误信息监听。
21）setOnVideoSizeChangedListener：视频尺寸监听。
22）setScreenOnWhilePlaying：设置是否使用 SurfaceHolder 来显示。
23）setVolume：设置音量。
24）start：开始播放。
25）stop：停止播放。
（3）使用注意事项。
1）新建（new）mediaplayer、reset( )方法后播放器处于空闲、使用 release( )方法后才会处于结束状态。
2）不使用 MediaPlayer 后，做好 release( )方法来释放，释放后就不能使用。
3）格式不支持、质量差时：通过 OnErrorListener.onError( )方法监控错误，用 reset( )方法恢复错误。
4）播放的顺序：先新建（或 reset），然后是 prepare，最后是 start。
5）一般在 start 前需要用 isplaying 方法检测。
6）pause：暂停。可用 start 恢复。
7）stop：停止。先调用 prepare( )准备，然后 start。
（4）举例：

```
mp=new MediaPlayer();
mp.setDataSource(路径);
mp.prepare();
mp.start();
```

2．音乐播放的实现

应用举例 Example_7_1——实现一个简单的 MP3 播放器功能，包含开始、暂停、结束、前一首、后一首功能。

创建应用程序，做好准备工作，将 MP3 播放器中要播放的音乐导入到模拟器的 SD 卡中。首先，设置模拟器 SD Card 的大小，以便用来存放歌曲，如图 7-1 所示。然后，启动模拟器，打开 DDMS，可以看到当前模拟器的文件目录，展开 mnt 目录，可以看到 sdcard 文件夹，该文件夹可于用于存放 mp3 音乐，如图 7-2 所示。最后，将所需音乐导入到该文件夹中，选中 mnt\sdcard 文件夹，单击导入按钮 ，选择想要播放的 MP3 音乐，音乐就会被导入到 SD 卡中。

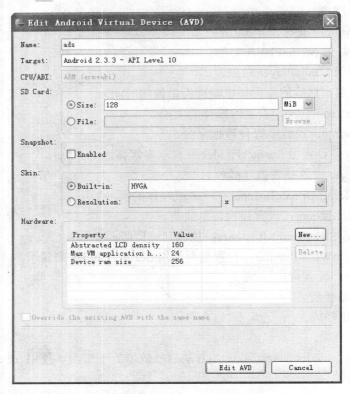

图 7-1　设置模拟器

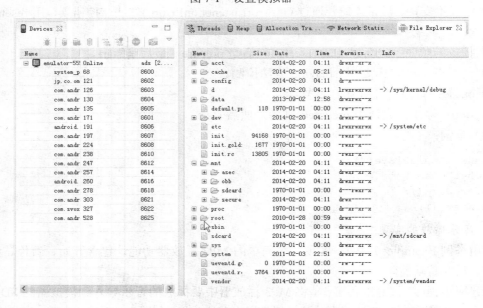

图 7-2　DDMS 中的文件目录

完成程序的界面设计：本程序用 ListView 控件制作播放音乐列表，将播放器按钮图片复制在 res/drawable 中，利用 5 个 ImageButton 控件做播放器按钮，本程序设置的主界面效果，如图 7-3 所示。

本程序源代码如下：

```java
package com.android;
import java.io.IOException;
import android.app.Activity;
import android.media.MediaPlayer;
import android.os.Bundle;
import android.view.View;
import android.view.View.OnClickListener;
import android.widget.AdapterView;
import android.widget.AdapterView.OnItemClickListener;
import android.widget.AdapterView.OnItemSelectedListener;
import android.widget.ArrayAdapter;
import android.widget.ImageButton;
import android.widget.ListView;
public class simplePlayer extends Activity {
    /** Called when the activity is first created. */
    //歌曲显示名称
    private String[] name=new String[]
                            {
        "爸爸妈妈","口是心非","谁是大英雄"
                            };
    //歌曲的 MP3 文件名称
    private String[] fiename=new String[]
                            {
        "babamama.mp3","ksxf.MP3","shuishidayingxiong.mp3"
                            };
    private ListView lv;
    private ImageButton ibLast,ibStart,ibPause,ibStop,ibNext;
    private MediaPlayer mp;
    private int i=0;
     @Override
     public void onCreate(Bundle savedInstanceState) {
        super.onCreate(savedInstanceState);
        setContentView(R.layout.main);
        lv=(ListView)findViewById(R.id.listView1);
        ibLast=(ImageButton)findViewById(R.id.imageButton1);
        ibStart=(ImageButton)findViewById(R.id.imageButton2);
        ibPause=(ImageButton)findViewById(R.id.imageButton3);
        ibStop=(ImageButton)findViewById(R.id.imageButton4);
        ibNext=(ImageButton)findViewById(R.id.imageButton5);
        mp=new MediaPlayer();//新建媒体
        ArrayAdapter<String> aa=new ArrayAdapter<String>(this, android.R.layout.simple_list_item_single_choice,name);
        lv.setAdapter(aa);
        lv.setOnItemClickListener(new listjt1());
```

图 7-3 简单 MP3 播放器的主界面

```java
        lv.setOnItemSelectedListener(new listjt2());
        ibLast.setOnClickListener(new jtLast());
        ibStart.setOnClickListener(new jtStart());
        ibPause.setOnClickListener(new jtPause());
        ibStop.setOnClickListener(new jtStop());
        ibNext.setOnClickListener(new jtNext());
    }
    //单击 ListView 中的项目时触发事件
    class listjt1 implements OnItemClickListener
    {
        public void onItemClick(AdapterView<?> arg0, View arg1, int arg2,
            long arg3) {
            i=arg2;
            playmusic();                           //调用播放音乐方法
        }
    }
    //ListView 项目改变时触发事件
    class listjt2 implements OnItemSelectedListener
    {
        public void onItemSelected(AdapterView<?> arg0, View arg1, int arg2,
            long arg3) {
            i=arg2;
        }
        public void onNothingSelected(AdapterView<?> arg0) {
            // TODO Auto-generated method stub
        }
    }
    //前一首
    class jtLast implements OnClickListener
    {
        public void onClick(View v) {
            // TODO Auto-generated method stub
            last();                                //调用上一首方法
            playmusic();                           //调用播放音乐方法
        }
    }
    //播放
    class jtStart implements OnClickListener
    {
        public void onClick(View v) {
            // TODO Auto-generated method stub
            playmusic();                           //调用播放音乐方法
        }
    }
    //暂停
    class jtPause implements OnClickListener
    {
        public void onClick(View v) {
            // TODO Auto-generated method stub
            if(mp.isPlaying())
            {
```

```
          mp.pause();                    //暂停播放
        }
        else
        {
          mp.start();                    //开始播放
        }
    }
  }
  //停止
  class jtStop implements OnClickListener
  {
    public void onClick(View v) {
      // TODO Auto-generated method stub
      if(mp.isPlaying())
      {
        mp.stop();                       //停止播放
      }
    }
  }
  //后一首
  class jtNext implements OnClickListener
  {
    public void onClick(View v) {
      // TODO Auto-generated method stub
      next();                            //调用下一首方法
      playmusic();                       //调用播放音乐方法
    }
  }
  //播放音乐方法
  public void playmusic()
  {
    mp.reset();                          //重置
    try {
      mp.setDataSource("/sdcard/"+fiename[i]);
                                         //设置媒体资源,找到SD卡中的文件
    } catch (IllegalArgumentException e) {
      e.printStackTrace();
    } catch (IllegalStateException e) {
      e.printStackTrace();
    } catch (IOException e) {
      e.printStackTrace();
        }
try {
    mp.prepare();                        //准备
    } catch (IllegalStateException e) {
      e.printStackTrace();
    } catch (IOException e) {
      e.printStackTrace();
    }
    mp.start();                          //开始
  }
```

```
   private void next()              //下一首方法,到达尾部从头开始
   {
      if(++i>=name.length)
      {
         i=0;
      }
   }
   private void last()              //上一首方法,到达首部从尾部开始
   {
      if(--i<=0)
      {
         i=name.length-1;
      }
   }
}
```

3. Video View 类

（1）作用：实现视频播放功能（包括 MP4 的 H.264、3GP 和 WMV 视频的解析）

（2）常用方法：

1）getBufferPercentage：得到缓冲的百分比。

2）getCurrentPosition：得到当前播放的位置。

3）getDuration：得到视频文件的时间。

4）isPlaying：是否正在播放。

5）pause：暂停。

6）reaolveAdjusetedSize：调整视频显示大小。

7）seekTo：指定播放位置。

8）setMediaController：设置播放控制器模式（播放进度条）。

9）setOnCompletionListener：当媒体文件播放完成时触发事件。

10）setOnErrorListener：错误监听。

11）setVideoPath：设置视频源路径。

12）setVideoURI 设置视频源地址。

13）start：开始播放。

（3）举例：

```
VideoView vv;
vv=(VideoView)findViewById(R.id.videoView1);
vv.setVideoPath("/sdcard/aa.mp4");
/* 设置模式-播放进度条*/
vv.setMediaController(new MediaController(video.this));
vv.requestFocus();
vv.start();
```

视频播放的实现。

应用举例 Example_7_2——实现一个视频播放器，包括装载、播放、暂停功能。

创建应用程序，做好准备工作，将视频播放器中要播放的视频文件导入到模拟器的 SD 卡中。

完成程序的界面设计：本程序用 3 个 Button 控件做播放器按钮，分别为装载、播放、暂停。本程序设置的主界面效果，如图 7-4 所示。

4. 本程序源代码如下：

图 7-4 视频播放器界面

```java
package com.yarin.android.Examples_07_03;
import com.yarin.android.Examples_07_03.R;
import android.app.Activity;
import android.os.Bundle;
import android.view.View;
import android.view.View.OnClickListener;
import android.widget.Button;
import android.widget.MediaController;
import android.widget.VideoView;
public class VideoPlayer extends Activity
{
    /** Called when the activity is first created. */
    @Override
    public void onCreate(Bundle savedInstanceState)
    {
        super.onCreate(savedInstanceState);
        setContentView(R.layout.main);
        /* 创建 VideoView 对象 */
        final VideoView videoView = (VideoView) findViewById(R.id.VideoView01);
        /* 操作播放的三个按钮 */
        Button PauseButton = (Button) this.findViewById(R.id.PauseButton);
        Button LoadButton = (Button) this.findViewById(R.id.LoadButton);
        Button PlayButton = (Button) this.findViewById(R.id.PlayButton);
        /* 装载按钮事件 */
        LoadButton.setOnClickListener(new OnClickListener()
        {
            public void onClick(View arg0)
            {
                /* 设置路径 */
                videoView.setVideoPath("/sdcard/aa.mp4");
                /* 设置模式——播放进度条 */
                videoView.setMediaController(new MediaController(VideoPlayer.this));
                videoView.requestFocus();
            }
        });
        /* 播放按钮事件 */
        PlayButton.setOnClickListener(new OnClickListener()
        {
            public void onClick(View arg0)
            {
                /* 开始播放 */
                videoView.start();
            }
        });
        /* 暂停按钮 */
```

```
    PauseButton.setOnClickListener(new OnClickListener()
    {
        public void onClick(View arg0)
        {
            /* 暂停 */
            videoView.pause();
        }
    });
  }
}
```

> **提 示**

在调试应用举例 Example_7_1 的过程中，用户可以发现，在音乐播放的过程中，再进行其他的操作，音乐一直在播放，但重新进入播放器时，不能对歌曲进行暂停、停止等操作。针对这个 BUG，要使用 Service 方法制作播放器。

7.2.2 Service 组件介绍

1. service 介绍

有些应用程序的应用，需要较长的执行时间（比如我们使用的 Alarmmanger），如果一直要保持这个应用程序的执行，势必会降低用户的使用。对于这些操作我们通常就采用 service 组件来实现。

用途：处理一些耗时比较长的操作，没有图形化界面。

举例：处理服务、发送 intent、启动系统通知。

2. service 的基本使用

启动 service：startService（intent）。

关闭 service：stopService（intent）。

使用步骤：

通过上面的方法启动或者关闭服务。

在 android 的 manifest 中增加服务。

`<service android:name=".服务名"></service>`

创建类并继承 service，复写相应的方法。

Service 相应的方法：

onbind( )：当要绑定 activity 时使用这个方法。

oncreate( )：当 service 初始化时调用这个方法。

onstartcommand( )：当创建一个 services 后会首先调用这个方法（主要功能实现）。

参数：intent 对象、flags、startid。

ondestory( )：当 service 被销毁时调用这个方法。

7.2.3 文本读取功能的实现

解决问题 1——文件的逐行读取。

（1）使用步骤：

1）获取文件信息：

```
File f=new File("/sdcard/a1.lrc");
```
2）将文件信息放入输入流：
```
FileInputStream fis=new FileInputStream(f);
```
3）读取输入流中的文件信息：
```
InputStreamReader isr=new InputStreamReader(fis);
```
4）将文件信息放入缓存中，实现逐行读取：
```
BufferedReader br=new BufferedReader(isr);
```
（2）BufferedReader 的用法。

readline( )：读出一行信息。

案例一：读取 lrc 文件中的一行信息。

案例二：读取出 lrc 文件中的所有信息。

解决问题 2——获得时间点。

对字符串操作的方法：

substring（开始的字符）。

substring（开始字符，结束字符）。

split（"条件"）：根据条件把字符串进行分割，结果保存到数组中。

案例一：读取时间点标志信息。

案例二：将时间信息转换成毫秒。

7.2.4 Handler 的使用

1．Handler 的基本概念

为程序提供了异步处理方案。

作用：可以进行另外线程的处理程序,优化程序。

应用举例：handler 实现每秒打印一次信息。

2．handler 的工作原理

线程队列：每个 handler 都会启动一个线程队列。

采用先进先出的处理机制，在程序运行中 handler 会不断的往这个队列中压入线程信息，根据压入的顺序一次执行这些线程信息。

3．handler 的基本使用方法

（1）post（线程）：立即向队列中压入线程。

（2）postDelayed（线程，毫秒）：通过指定的毫秒数向队列中压入线程。

（3）removeCallbacks（线程）：注销 handler 的线程队列。

1）创建一个 Handler 对象：
```
Handler handler = new Handler();
```
2）主程序中通过 handler 执行线程的调用：
```
Handler.post(updateThread);
```
3）将要执行的操作写入线程对象的 run 方法中去：
```
Runnable updateThread = new Runnable (){
```

```
Public void run() {
System.out.println("UpdateThread");
handler.postDelayed(updateThread , 3000);        // 延迟加入消息队列
}}
```

案例：实现每 3 秒，打印一行 lrc 文件中的歌词。

4. 歌词同步问题的设计与实现

实现思想：

1）将 lrc 歌词进行分解：时间，歌词。

2）将时间信息转换成毫秒后存储动态数组，歌词存入动态数组。

3）通过动态数组中第一个时间，计算调用 handler 的时间。

4）执行线程时显示歌词，在线程中计算下一次调用 handler 的时间。

案例：在界面上添加：

1）开始、结束按钮并显示歌词控件。

2）使用 service 实现播放 MP3 功能。

3）通过 handler 实现歌词同步的功能。

7.3 任务实施

7.3.1 任务描述

实现完整的媒体播放器功能。

7.3.2 实现步骤

（1）准备开发用资源（程序中）和测试用资源（sdcard 下）。

（2）界面设计。

（3）在后台进行映射控件。

（4）保证界面能够正常显示需要的信息（此程序实现在 listview 上绑定歌名）。

1）数据准备：把歌名写入动态数组，同时把对应 lrc 也写入另外动态数组。

注意 lrc 存储时，默认所有歌都带歌词，用歌名来写 lrc 动态数组。

2）准备适配器：

3）数据绑定：

（5）做第一次测试，保证界面和 lrc 有信息。

（6）实现基本的 MP3 播放功能，在 servie 下实现（传递歌名路径+操作状态）。

1）Zt 1 播放   zt 2 暂停   zt 3 停止

2）创建类并集成 service。

3）注册 service。

（7）完成各个事件的监听处理。

（8）写各个方法的注释。

（9）设置统一的调用 service 方法。

（10）实现各个监听器调用 9 中的方法。

（11）进行第二次测试，保证 MP3 基本能够使用。

(12) 歌词同步问题的实现。

1) 写一个方法实现 gc 同步：判定是否有该 lrc（exits 方法），将 lrc 信息分解读入对应动态数组（注意时间信息要进行转制成 Long）。

2) 建立线程实现 textview 显示歌词。

3) 用 handler 进行线程间消息传递。

(13) 进行整体测试。

1) 界面 1：（发起）。

①创建 Intent 对象；

②设置传递的信息，使用 putExtra（键值名，键值）方法；

③跳转的规则，使用 setClass（界面 1 类，界面 2 类）方法；

④跳转，使用 startActivity（Intent 对象）。

2) 界面 2：（接收）。

①获得 Intent，使用 getIntent( )方法；

②读取并使用信息，使用 getExtra（键值名）方法。

7.3.3 界面设计

设计 MP3 播放器界面，上方使用进度条显示播放器的播放进度，中间显示要播放的歌曲，下方是歌词同步显示，具体界面如图 7-5 所示。

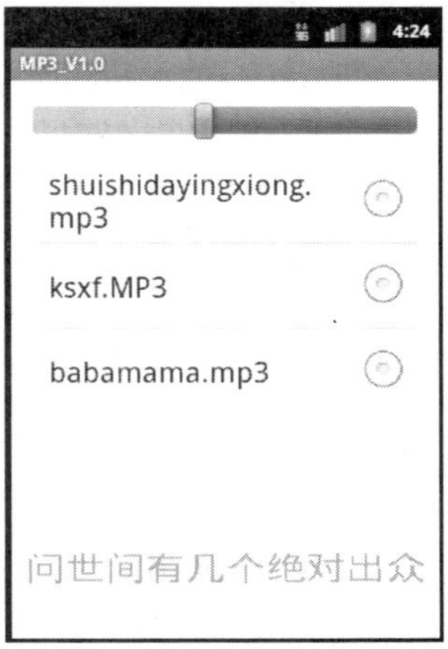

图 7-5 MP3 播放器界面

7.3.4 代码实现

（1）MainActivity.java 文件。

```
package com.example.mp3;

import java.io.File;
import java.util.ArrayList;

import android.os.Bundle;
import android.app.Activity;
import android.content.BroadcastReceiver;
import android.content.Context;
import android.content.Intent;
import android.content.IntentFilter;
import android.graphics.Bitmap;
import android.graphics.Bitmap.Config;
import android.graphics.Canvas;
import android.graphics.Color;
import android.graphics.Paint;
import android.graphics.drawable.BitmapDrawable;
import android.graphics.drawable.Drawable;
import android.util.DisplayMetrics;
import android.view.Menu;
import android.view.MenuItem;
import android.view.View;
import android.widget.AdapterView;
import android.widget.AdapterView.OnItemClickListener;
```

```java
import android.widget.ArrayAdapter;
import android.widget.ListView;
import android.widget.RelativeLayout;
import android.widget.SeekBar;
import android.widget.Toast;
public class MainActivity extends Activity {
    //准备控件对应的对象
    ListView gmlist;                                             //显示歌名的列表
    SeekBar  gprogress;                                          //显示歌曲的进度
    BroadcastReceiver broad,broad1;                              //广播接收器
    RelativeLayout   rl;                                         //界面的布局样式

    ArrayList<String> namelist=new ArrayList<String>();  //所有歌名,动态数组
        protected void onCreate(Bundle savedInstanceState) {
        super.onCreate(savedInstanceState);
        setContentView(R.layout.activity_main);
        //查找界面上对应控件
        gmlist=(ListView) findViewById(R.id.listView1);
        gprogress=(SeekBar) findViewById(R.id.seekBar1);
        rl=(RelativeLayout) findViewById(R.id.relative1);
        //初始化工作:初始化界面,显示歌曲的列表
        getgmlist();
        //实现单击列表播放歌曲
        gmlist.setOnItemClickListener(new bflistener());
    }
    class bflistener implements OnItemClickListener
    {
        @Override
        public void onItemClick(AdapterView<?> arg0, View arg1, int arg2,
            long arg3) {
        // 处理单击播放:用 service 完成,主界面下就要传递信息给 service
            //此处要像 service 中传递状态和歌曲的位置
            Intent t=new Intent();
            t.setClass(MainActivity.this, mp3service.class);
            t.putExtra("mp3zt", 1);//1 表示播放,2 表示停止
            String path="/sdcard/"+namelist.get(arg2);           //具体播放歌曲的路径
            t.putExtra("mp3path", path);                         //传递歌曲的路径
            startService(t);
        }
    }
    private void getgmlist() {
        // 获取 sdcard 下面 MP3 文件的列表,采用文件操作
        File root=new File("/sdcard/");                          //获得 sdcrad 的根目录
        File[] rootlist=root.listFiles();           //搜索 sdcrad 目录下所有文件
        for(int i=0;i<rootlist.length;i++)
        {
            //搜索所有获得的文件,通过后缀过滤非 MP3 文件 if(rootlist[i].getName().endsWith(".mp3")||rootlist[i].getName().endsWith(".MP3"))
            {
                //当后缀满足文件类型要求时,则像歌曲名列表中增加该文件
                namelist.add(rootlist[i].getName());
```

```java
            }
        }
        //将存在的歌名信息,绑定到listview控件中
        ArrayAdapter<String> adapter=new
   ArrayAdapter<String>(MainActivity.this, android.R.layout.select_dialog_singlechoice, namelist);
        gmlist.setAdapter(adapter);
    }
    @Override
    public boolean onCreateOptionsMenu(Menu menu) {
        // Inflate the menu; this adds items to the action bar if it is present.
        getMenuInflater().inflate(R.menu.main, menu);
        return true;
    }
    @Override
    public boolean onOptionsItemSelected(MenuItem item) {
        // 判定是否单击停止按钮
        if(item.getItemId()==R.id.stop)
        {
            Intent t=new Intent();
            t.setClass(MainActivity.this, mp3service.class);
            t.putExtra("mp3zt", 2);                    //1表示播放,2表示停止
            t.putExtra("mp3path", "");                 //传递歌曲的路径
            startService(t);
        }
        return super.onOptionsItemSelected(item);
    }
    @Override
    protected void onPause() {
        // 当程序退出时,注销广播
        super.onPause();
        //注销广播
        unregisterReceiver(broad);
        unregisterReceiver(broad1);
    }

    @Override
    protected void onResume() {
        // 当程序进入时,注册广播
        super.onResume();
        //注册广播:接收器、频率
        broad=new gcgb();
        broad1=new seekbargb();
        registerReceiver(broad, getintentFilter());
        registerReceiver(broad1, getintentFilter1());
    }
    IntentFilter getintentFilter()
    {
        //规定接收器的频率
        IntentFilter infilter=new IntentFilter();
        infilter.addAction("musiclrc");
```

```java
            return infilter;
        }
        IntentFilter getintentFilter1()
        {
            //规定接收器的频率
            IntentFilter infilter=new IntentFilter();
            infilter.addAction("seekbartb");
            return infilter;
        }

        class gcgb extends BroadcastReceiver
        {
            //接收线程中的广播
            public void onReceive(Context context, Intent intent) {
                //接收广播的内容
                String  lrc=intent.getStringExtra("lrc");
                //画歌词
                drawlrc(lrc);
            }
            private void drawlrc(String lrc) {
                //画歌词:准备bitmap 在上面画歌词,将带歌词的bitmap设置为界面的背景
                DisplayMetrics dm=new DisplayMetrics();
                getWindowManager().getDefaultDisplay().getMetrics(dm);
                //准备图片
                Bitmap bmp=Bitmap.createBitmap(dm.widthPixels, dm.heightPixels, Config.ARGB_8888);
                //以该图片为背景准备画布
                Canvas c=new Canvas(bmp);
                //准备画笔
                Paint p=new Paint();
                p.setColor(Color.GREEN);
                p.setTextSize(30);
                //用画笔在画布上画文字
                c.drawText(lrc, 10, dm.heightPixels-50, p);
                //准备drawable 文件
                Drawable draw=new BitmapDrawable(bmp);
                //以该drawable 文件为界面背景
                rl.setBackgroundDrawable(draw);
            }

        }
        class seekbargb extends BroadcastReceiver
        {

            @Override
            public void onReceive(Context context, Intent intent) {
                // TODO Auto-generated method stub
                int wz=intent.getIntExtra("wz", 0);
                int zlong=intent.getIntExtra("long",0);
                gprogress.setMax(zlong);
                gprogress.setProgress(wz);
```

            }
        }
    }

（2）mp3service.java 文件。

```java
package com.example.mp3;
import java.io.BufferedReader;
import java.io.File;
import java.io.FileInputStream;
import java.io.FileNotFoundException;
import java.io.IOException;
import java.io.InputStreamReader;
import java.io.UnsupportedEncodingException;
import java.util.ArrayList;
import android.app.Service;
import android.content.Intent;
import android.media.MediaPlayer;
import android.os.Handler;
import android.os.IBinder;
import android.provider.MediaStore.Audio.Media;
import android.widget.Toast;
public class mp3service extends Service {
    MediaPlayer media=new MediaPlayer();
    String lrc=null;//存储某句歌词
    ArrayList<String> gclist=new ArrayList<String>();    //存储歌词信息
    ArrayList<Long> sjlist=new ArrayList<Long>();        //存储歌词时间信息
    int i=0;                                             //控制歌词所在的位置
    public IBinder onBind(Intent intent) {
        // TODO Auto-generated method stub
        return null;
    }
    @Override
    public int onStartCommand(Intent intent, int flags, int startId) {
        // 接收MP3信息,实现音乐播放器的播放
        int mp3zt=intent.getIntExtra("mp3zt", 0);
        String path=intent.getStringExtra("mp3path");
        System.out.println(path);
        switch(mp3zt)
        {
        //判定不同的状态,1进行播放,2停止播放
        case 1:playmusic(path);break;
        case 2:stopmusic();break;
        }
        return super.onStartCommand(intent, flags, startId);
    }
    private void stopmusic() {
        // TODO Auto-generated method stub
        media.stop();
        h.removeCallbacks(gcxs);
        //media.release();
    }
```

```java
    private void playmusic(String path) {
        //寻找对应 lrc 文件
        File gcfile=null;
        if(path.endsWith(".mp3"))
        {
//          //判定后缀为 MP3 的文件,用 lrc 进行替换
            gcfile=new File(path.replace(".mp3", ".lrc"));
        }
        if(path.endsWith(".MP3"))
        {
            gcfile=new File(path.replace(".MP3", ".lrc"));
        }
        if(gcfile.exists())
        {
            h.removeCallbacks(gcxs);
            h.removeCallbacks(seekbarxs);
            //如果歌词文件存在,获取歌词文件中的时间和具体歌词
            preparelrc(gcfile);
            i=0;
            h.postDelayed(gcxs, sjlist.get(i));
            hh.post(seekbarxs);
        }
        // 播放歌曲
        media.reset();
        try {
            media.setDataSource(path);
            media.prepare();
        } catch (IllegalArgumentException e) {
            // TODO Auto-generated catch block
            e.printStackTrace();
        } catch (IllegalStateException e) {
            // TODO Auto-generated catch block
            e.printStackTrace();
        } catch (IOException e) {
            // TODO Auto-generated catch block
            e.printStackTrace();
        }
        media.start();
    }
    private void preparelrc(File gcfile) {
        //拆解 lrc 文件获取时间和歌词,写入对应数组中
        //JAVA IO 操作
        //1) 先把文件输入信息流
        FileInputStream fis;
        try {
            fis=new FileInputStream(gcfile);
            //2) 读取输入信息流,注意码制转换
            InputStreamReader isr=new InputStreamReader(fis,"gb2312");
            //3) 逐行解析 lrc 文件,注意清空原有的歌词和时间
            gclist.clear();
            sjlist.clear();
```

```java
                BufferedReader br=new BufferedReader(isr);
                String xx;
                while((xx=br.readLine())!=null)
                {
                    //根据歌词样式获取lrc中歌词信息
                    gclist.add(xx.substring(10));
                    //将lrc中时间进行转换,以毫秒为单位进行存储
                    Long msj=zhhm(xx.substring(1, 9));
                    sjlist.add(msj);
                }
            } catch (FileNotFoundException e) {
                // TODO Auto-generated catch block
                e.printStackTrace();
            } catch (UnsupportedEncodingException e) {
                // TODO Auto-generated catch block
                e.printStackTrace();
            } catch (IOException e) {
                // TODO Auto-generated catch block
                e.printStackTrace();
            }
        }
        private Long zhhm(String substring) {
            // 将分、秒、毫秒 转换成毫秒
            String[]  a=substring.split(":");
            String[]  b=a[1].split("\\.");
            //切割后a[0]存储分钟,b[0]存储秒钟,b[1]存储毫秒,运算结果以毫秒为单位
            Long msj=
                (long) (Integer.parseInt(a[0])*60*1000
                    +Integer.parseInt(b[0])*1000
                    +Integer.parseInt(b[1])*10);
            return msj;
        }
        Handler h=new Handler();
        Runnable gcxs=new Runnable() {
            @Override
            public void run() {
                //歌词同步:1)取当前一句歌词;2)以广播形式发送界面,界面显示歌词;3)到下个时刻显示下一句歌词
                lrc=gclist.get(i);
                //广播发送当前歌词
                Intent t=new Intent();
                t.putExtra("lrc", lrc);
                t.setAction("musiclrc");
                sendBroadcast(t);
                //准备下一句歌词
                long nowtime=sjlist.get(i);
                i++;
                if(gclist.get(i)!=null)
                {
                    //如果存在下一句歌词,调用该线程
                    long nexttime=sjlist.get(i);
```

```
            long waittime=nexttime-nowtime-20;
            h.postDelayed(gcxs, waittime);
        }
    }
};
Handler hh=new Handler();
Runnable seekbarxs=new Runnable() {
    @Override
    public void run() {
        // 设定一个间隔,不断发送当前 MP3 位置
        int wz=media.getCurrentPosition();
        int zlong=media.getDuration();
        Intent t=new Intent();
        t.putExtra("wz", wz);
        t.putExtra("long", zlong);
        t.setAction("seekbartb");
        sendBroadcast(t);
        hh.postDelayed(seekbarxs, 500);
    }
};
}
```

第 8 章 Android 游戏开发

▲ 引言

目前在中国移动互联网时代，兴起了安卓热，安卓手机、平板等移动智能终端的市场不断增长。现在的孩子们，人手一台移动终端设备不足为奇，并且，对愤怒的小鸟、tom 猫、水果忍者等游戏甚是喜爱。就是因为喜欢那些在手机上安装的好玩游戏，喜欢移动中打游戏的时刻，出于强烈的好奇心，很多人对游戏开发很感兴趣。

8.1 需求分析

8.1.1 利用 Android 制作打地鼠游戏

打地鼠是一个趣味性的休闲游戏，在新游戏页面单击游戏图标，即可直接进入游戏。此刻，地鼠会从一个个地洞中不经意的探出一个脑袋，地鼠冒出来后左键点会出现一个大锤子敲在地鼠头上，打中加 1 分。

8.1.2 利用 Android 制作俄罗斯方块游戏

俄罗斯方块是一款风靡全球的手机游戏，它由俄罗斯人阿列克谢·帕基特诺夫发明，故得此名。俄罗斯方块的基本规则是移动、旋转和摆放游戏自动输出的各种方块，使之排列成完整的一行或多行并且消除得分。由于上手简单，老少皆宜，从而家喻户晓，风靡世界。

8.2 知识准备

8.2.1 View 类开发框架

1. 主要技术点

（1）在 view 类中使用 onDraw 实现界面。

（2）在 View 类中交互事件：onKeyUp、onKeyDown、onTouchEvent。

（3）界面刷新：invalidate 或 post invalidate（可放在线程中使用）。

2. onTouchEvent 的基本使用（单击时可以获得单击点的坐标，通过这个坐标完成相关事件的处理）

（1）getx( )、gety( )：获得单击事件的坐标。

（2）getaction( )：获得事件对应动作。

（3）action_down,action_up：单击屏幕按下和抬起操作。

案例：单击屏幕，获得单击点坐标，使用 toast 进行显示。

```
public boolean onTouchEvent(MotionEvent event) {
    float x=event.getX();
    float y=event.getY();
    Toast.makeText(sy.this,"坐标:"+x+","+y, Toast.LENGTH_SHORT).show();
    return true;
}
```

8.2.2 绘制矩形的知识点

游戏需要不断的在屏幕上绘制图形——引入 Graphics 类的使用。

包括：Canvas（画布）、Paint（画笔）、Color（颜色）、Bitmap（图像）。

1. 绘图步骤

（1）调整画笔。

（2）将图像绘制到画布上。

2. Paint 类

常用的方法：

（1）setColor：设置画笔的颜色。

（2）setAlpha：设置透明度。

（3）setTextSize：设置字体尺寸。

（4）setStyle：设置画笔的风格（空心、实心）。

Style.STROKE：空心。

Style.FILL：实心。

（5）setStrokeWidth：设置空心的边框宽度。

案例一：

画一个矩形要求：200*100，颜色为红色，半透明，空心，边宽为 5。

代码实现：

（1）新建一个 view 类。

```
public class view1 extends View{
    private Paint p;//声明画笔对象
    public view1(Context context) {
        super(context);
        p=new Paint();//对画笔对象实例化
    }
    protected void onDraw(Canvas canvas) {
        super.onDraw(canvas);
        p.setColor(Color.RED);              //设置画笔颜色为红色
        p.setAlpha(50);                     //设置透明度
        p.setStyle(Style.STROKE);           //设置空心
        p.setStrokeWidth(5);                //设置边框宽度
        canvas.drawRect(0, 0, 200, 100, p); //画矩阵形 200*100
    }
}
```

（2）在 Activity 中调用 view 类。

```
public class MainActivity extends Activity {
```

```
    public void onCreate(Bundle savedInstanceState) {
        super.onCreate(savedInstanceState);
        view1 v;
        v=new view1(MainActivity.this);
        setContentView(v);
    }
    public boolean onCreateOptionsMenu(Menu menu) {
        getMenuInflater().inflate(R.menu.activity_main, menu);
        return true;
    }
}
```

案例二：

画一个矩形要求：100*200，颜色为蓝色，不透明，实心，边宽为1。

```
public class view2 extends View{
    private Paint p;                              //声明画笔对象
    public view1(Context context) {
        super(context);
        p=new Paint();                            //对画笔对象实例化
    }
    @Override
    protected void onDraw(Canvas canvas) {
        super.onDraw(canvas);
        p.setColor(Color.BLUE);                   //设置画笔颜色为蓝色
        p.setAlpha(100);                          //设置透明度为不透明
        p.setStyle(Style.FILL);                   //设置实心
        p.setStrokeWidth(1);                      //设置边框宽度
        canvas.drawRect(0, 0, 100, 200, p);       //画矩阵形100*200
    }
}
```

3. Canvas 类

常用方法：

drawColor：设置画布的背景颜色。

Canvas( )：创建一个空的画布。

案例三：

在案例二基础上设置一个白色的画布。

```
canvas.drawColor(Color.WHITE);
```

4. 几何图形的绘制

常用方法：

drawRect：绘制矩形（左上坐标，右下坐标，画笔）。

drawCircle：绘制圆形（圆心坐标，半径，画笔）。

drawLine：绘制直线。

drawPoint：绘制点。

案例四：

在屏幕上绘制一个半径为 50 的圆。

```java
public class view4 extends View{
    private Paint p;                                //声明画笔对象
    public view1(Context context) {
        super(context);
        p=new Paint();                              //对画笔对象实例化
    }
    @Override
    protected void onDraw(Canvas canvas) {
        super.onDraw(canvas);
        p.setColor(Color.BLUE);                     //设置画笔颜色为蓝色
        canvas.drawCircle(100, 100, 50, p);//以(100,100)为中心画半径为 50 的圆
    }
}
```

案例五：

在屏幕上绘制一条直线。

```java
public class view5 extends View{
    private Paint p;                                //声明画笔对象
    public view5(Context context) {
        super(context);
        p=new Paint();                              //对画笔对象实例化
    }
    @Override
    protected void onDraw(Canvas canvas) {
        super.onDraw(canvas);
        p.setColor(Color.BLUE);                     //设置画笔颜色为蓝色
        canvas.drawLine(200, 200, 300, 300, p);     //绘制直线起点是(200,200),终点
                                                    //是 (300,300)
    }
}
```

8.2.3 字符串的绘制

常用方法：

drawText 方法（字符串、开始坐标、pen）。

案例六：在屏幕上输入："你好"，字体为 30 号字，颜色为黑色。

```java
public class view6 extends View{
    private Paint p;                                //声明画笔对象
    public view6(Context context) {
        super(context);
        p=new Paint();                              //对画笔对象实例化
    }
    @Override
    protected void onDraw(Canvas canvas) {
        super.onDraw(canvas);
        p.setColor(Color.BLACK);                    //设置画笔颜色为黑色
        p.setTextSize(30);                          //设置字号为 30
        canvas.drawText("hello world", 50, 50, p);  //在(50,50)位置写字符串hello world
    }
```

}

8.2.4 图像的绘制

常用方法：

drawBitmap（bitmap 对象，起点坐标，null）；

如果要使用 drawable 资源，需要将资源转换成 Bitmap 对象方法如下：

Bitmap bi=((BitmapDrawable)getResources( ).getDrawable(id 信息)).getBitmap( );

图像的绘制：

创建一个新的位图，使用原有的位图。

createBitmap(原位图,原图坐上坐标（0，0），原图右下坐标, matrix,true);

Matrix 图片的操作（postScale（x,y）按 x,y 的比例进行图的缩放）

案例七：

将程序默认的 icon 图绘制到界面上。

```
public class view7 extends View{
    private Paint p;                    //声明画笔对象
    public view7(Context context) {
       super(context);
       p=new Paint();                   //对画笔对象实例化
    }
    @Override
    protected void onDraw(Canvas canvas) {
       super.onDraw(canvas);
        Bitmap bi=((BitmapDrawable)getResources().getDrawable(R.drawable.ic_launcher)).getBitmap();
         canvas.drawBitmap(bi, 100,100,p);
    }
}
```

8.3 任务实施（一）

8.3.1 任务描述

打地鼠规则设定：

生命设置为 20（初始值生命 20）

爆机进度达到 800（进度的初始值 0）

当打击到地鼠，进度+1；

打地鼠缩回洞中之前并没有打击到时，生命−1；

初始时，所有洞都显示为空（开始点坐标从 35,60 开始，每个图片的尺寸 80*80，行数为 4，列数为 3，洞数 12）。

当打击到地鼠时显示打击成功的图片。

8.3.2 实现步骤

（1）建立一个打地鼠的结构框架。

1）建立一个游戏入口（activity）。

2）建立一个界面类（继承 view）。

3）控制类，主要控制动画的一个播放。

（2）初始化游戏的规则，以及做准备工作（生命（20）、分数（0）、行数（4）、列数（3）、洞数（12）、每个洞的尺寸（80））。

（3）完成动画的设计：

1）地鼠出洞动画。

2）地鼠被打到动画。

（4）完成动画控制的编写。

1）设置洞相关的状态（初始状态、播放状态、打击状态、实时状态）。

2）定义动画的各种方法：初始动画、打击动画、下一时刻动画。

（5）对每一个洞洞进行初始化：使用动态链表来实现。

（6）开始绘制游戏界面：绘制背景、绘制文字、通过每个洞的状态绘制每个洞的 bitmap（绘制结束后要调用下一时刻的状态）。

（7）实现界面的刷新（每 100 毫秒刷新一次）。

（8）实现每一秒钟出现一只地鼠。

1）首先判定洞是否为空，如果为空则将该洞加入到临时列表中。

2）随机临时列表中的所有可能的洞，使其中一个洞播放动画。

（9）触发触摸屏事件，当按下触屏时，则获取 x、y 坐标还原成初始坐标，再判定是否击中，击中则播放打击动画。

（10）事件处理，在动画播放结束之前如果没有打倒生命--，如果在播放动画过程中打击到事件则分数++。

（11）设置程序结束方法（当生命为 0 分数大于 800 时在刷新线程中关闭线程）。

8.3.3　界面设计

打地鼠游戏界面主要包括地洞和地鼠，还有生命和得分，游戏结束后，会弹出对话框，可以重新开始或退出游戏。具体界面如图 8-1 所示。

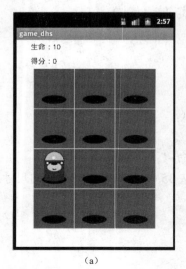　　

（a）　　　　　　　　　　　　　　　（b）

图 8-1　打地鼠游戏界面

（a）游戏主界面；（b）游戏结束界面

8.3.4 代码实现

(1) Main_dds.java 文件。

```java
package com.example.game_dhs;
import android.os.Bundle;
import android.app.Activity;
import android.util.DisplayMetrics;
import android.view.Menu;
public class Main_dds extends Activity {
    //程序入口:对于不同类的调用,游戏界面的引入
    Dds_view dv;
    //定义窗体的长宽
    int windows_width;
    int windows_height;
    protected void onCreate(Bundle savedInstanceState) {
        super.onCreate(savedInstanceState);

        startGame();
    }
    public void startGame() {
        // TODO Auto-generated method stub
        //得到当前窗体的宽高
        DisplayMetrics dm=new DisplayMetrics();
        //用DM存储窗体初始化信息
        getWindowManager().getDefaultDisplay().getMetrics(dm);
        //分别取出当前窗体的宽高
        windows_width=dm.widthPixels;
        windows_height=dm.heightPixels;
        dv=new Dds_view(Main_dds.this,windows_width,windows_height);
        setContentView(dv);
    }
    @Override
    public boolean onCreateOptionsMenu(Menu menu) {
        // Inflate the menu; this adds items to the action bar if it is present.
        getMenuInflater().inflate(R.menu.main_dds, menu);
        return true;
    }
}
```

(2) Dds_view 文件。

```java
package com.example.game_dhs;
import java.util.ArrayList;
import java.util.HashMap;
import android.app.AlertDialog;
import android.content.Context;
import android.content.DialogInterface;
import android.graphics.Bitmap;
import android.graphics.Canvas;
import android.graphics.Color;
import android.graphics.Matrix;
import android.graphics.Paint;
```

```java
import android.graphics.drawable.BitmapDrawable;
import android.os.Handler;
import android.util.MonthDisplayHelper;
import android.view.MotionEvent;
import android.view.View;
public class Dds_view extends View {
    //定义动画对应hashmap
    HashMap<Integer, Bitmap> values=new HashMap<Integer, Bitmap>();
    //线性链表
    ArrayList<Pic> hole=new ArrayList<Pic>();
    //每个地鼠的大小
    int SIZE;
    int life=20;
    int score=0;
    int startx=35;
    int starty=60;
    AlertDialog.Builder builder;
    public Dds_view(Context context,int win_width,int win_height) {
        super(context);
        //初始化每一个洞的大小
        SIZE=(win_width-startx*2)/3;
        // 图形的初始化:两组动画 :正常地鼠动画,打中地鼠的动画
        //1)正常地鼠动画:规范图片,确定图片在程序的大小
        values.put(1, tpzh(R.drawable.show1));
        values.put(2, tpzh(R.drawable.show2));
        values.put(3, tpzh(R.drawable.show3));
        values.put(4, tpzh(R.drawable.show4));
        values.put(5, tpzh(R.drawable.show5));
        values.put(6, tpzh(R.drawable.show6));
        values.put(7, tpzh(R.drawable.show6));
        values.put(8, tpzh(R.drawable.show6));
        values.put(9, tpzh(R.drawable.show5));
        values.put(10, tpzh(R.drawable.show4));
        values.put(11, tpzh(R.drawable.show3));
        values.put(12, tpzh(R.drawable.show2));
        values.put(13, tpzh(R.drawable.show1));
        //2)没地鼠的样子
        values.put(0, tpzh(R.drawable.emptyhole));
        //3)打中地鼠动画
        values.put(-9, tpzh(R.drawable.hit));
        values.put(-8, tpzh(R.drawable.hit));
        values.put(-7, tpzh(R.drawable.hit));
        values.put(-6, tpzh(R.drawable.hit));
        values.put(-5, tpzh(R.drawable.show5));
        values.put(-4, tpzh(R.drawable.show4));
        values.put(-3, tpzh(R.drawable.show3));
        values.put(-2, tpzh(R.drawable.show2));
        values.put(-1, tpzh(R.drawable.show1));
        //初始化每一个洞,相当于初始化线性链表
        for(int i=0;i<Constants.HOLE_COUNT;i++)
        {
```

```java
            hole.add(new Pic());
        }
        hh.postDelayed(sxscreen, 80);         //每隔80毫秒刷新一次屏幕
        h.postDelayed(sxds,1000);              //每隔1000毫秒产生一只地鼠
    }
    Handler hh=new Handler();
    //屏幕刷新问题：定时刷新屏幕,设置间隔较短
    Runnable sxscreen =new Runnable() {
        @Override
        public void run() {
            // 定时刷新屏幕,80毫秒
            postInvalidate();
            hh.postDelayed(sxscreen, 80);
        }
    };
    Handler h=new Handler();
    //实现每一秒钟产生新的地鼠
    Runnable sxds=new Runnable() {
        @Override
        public void run() {
            // 每间隔一秒钟产生一个新的地鼠
            //产生一个链表,存储当前空着的洞
            ArrayList<Pic> temp=new ArrayList<Pic>();
            for(int i=0;i<hole.size();i++)
            {
                //判定某一个洞为空时,存储这个空的洞
                if(hole.get(i).now_zt==0)
                {
                    //当某一个洞为空时,写入到temp链表
                    temp.add(hole.get(i));
                }
            }
            //定义n 表示有几个洞空着
            int n=temp.size();
            if(n==0)
            {
                //占满时的操作
                h.removeCallbacks(sxds);
                hh.removeCallbacks(sxscreen);
            }
            else
            {
                //产生地鼠
                int ram=(int) (Math.random()*n);
                temp.get(ram).thShow();
                h.postDelayed(sxds,1000);
            }
        }
    };
    private Bitmap tpzh(int resid) {
        // 将图片进行等比例缩放
```

```java
            //先获取资源图片信息
            Bitmap y_bitmap=((BitmapDrawable)(getResources().getDrawable(resid))).getBitmap();
            //获取原始图片的长宽信息
            int y_width=y_bitmap.getWidth();
            int y_height=y_bitmap.getHeight();
            //等比例缩放
            Matrix matrix=new Matrix();
            //设置缩放的比例 s_width  s_height
            float s_width=(float)SIZE/y_width;
            float s_height=(float)SIZE/y_height;
            //设置缩放比例
            matrix.postScale(s_width, s_height);
            //利用比例信息缩放原始图片
            Bitmap x_bitmap=Bitmap.createBitmap(y_bitmap, 0, 0, y_width, y_height, matrix, true);
            return x_bitmap;
        }
        //游戏控制
        public boolean onTouchEvent(MotionEvent event) {
            // 地鼠打击判定
            //判定已经单击了,则判定是否打中地鼠
            if(event.getAction()==MotionEvent.ACTION_DOWN)
            {
                //判定是否打在范围内
                int x=(int) ((event.getX()-startx)/SIZE);
                int y=(int) ((event.getY()-starty)/SIZE);
                if(x<0||x>=3||y>=4||y<0)
                {
                    //打到外面去了
                    return true;
                }
                else
                {
                    //打到里面的操作
                    //判定是那个洞:0-11 第几个洞:y*3+x
                    if(hole.get(y*3+x).hit_pd())
                    {
                        //打中了
                        score++;
                    }
                    return true;
                }
            }
            return true;
        }
        protected void onDraw(Canvas canvas) {
            //画图信息
            if(life<=0)
            {
                //游戏结束:撤销进程
```

```java
            h.removeCallbacks(sxds);
            hh.removeCallbacks(sxscreen);
            //游戏结束操作
            doGameover();
        }
        super.onDraw(canvas);
        canvas.drawColor(Color.WHITE);
        //准备画笔
        Paint p=new Paint();
        p.setTextSize(15);
        p.setColor(Color.BLUE);
        //画生命
        canvas.drawText("生命:"+life, 29, 20, p);
        //画得分
        canvas.drawText("得分:"+score, 29, 50, p);
        //画洞 :当前洞的信息
        for(int i=0;i<hole.size();i++)
        {
            Pic pic=hole.get(i);
            //确定每一个洞所在的行列
            //y 表示第几行
            int y=i/Constants.COLUMN_COUNT;
            //x 表示某一列
            int x=i%Constants.COLUMN_COUNT;
            //确定某一时刻各个洞中图的样子
            Bitmap now_bm=values.get(pic.now_zt);
            //画图:注意坐标
            canvas.drawBitmap(now_bm, startx+x*SIZE, starty+y*SIZE, null);
            //当前状态播放结束后
            if(pic.toNext())
            {
                life--;
            }
        }
    }
    private void doGameover() {
        // 游戏结束:弹出窗体告知游戏结束
        builder=new AlertDialog.Builder(getContext());
        //设置窗体标题
        builder.setTitle("游戏结束啦！");
        //设置窗体的内容
        builder.setMessage("单击开始新的游戏");
        //设置背景不可单击
        builder.setCancelable(false);
        //重新开始按钮设置
        builder.setPositiveButton("重新开始", new jt1());
        //退出按钮的设置
        builder.setNegativeButton("退出游戏", new jt2());
        //展示按钮
        builder.show();
    }
```

```java
class jt1 implements android.content.DialogInterface.OnClickListener
{
    @Override
    public void onClick(DialogInterface dialog, int which) {
        // 重新开始游戏操作
        Main_dds md=(Main_dds) getContext();
        md.startGame();
        builder.setCancelable(true);
    }
}
class jt2 implements android.content.DialogInterface.OnClickListener
{
    @Override
    public void onClick(DialogInterface dialog, int which) {
        // 退出游戏操作
        android.os.Process.killProcess(android.os.Process.myPid());
    }
}
}
```

（3）Constants.java。

```java
package com.example.game_dhs;

public interface Constants {

    //本应用程序的一些常量信息:恒定不变的量
    //列数
    int COLUMN_COUNT=3;
    //行数
    int ROW_COUNT=4;
    //洞数
    int HOLE_COUNT=COLUMN_COUNT*ROW_COUNT;
}
```

（4）Pic.java。

```java
package com.example.game_dhs;

public class Pic {
    //各种图的信息处理①正常的画面；②打中的画面；③打中的判定；④打不中的判定；⑤出现地鼠
    int cs_information=0;                       //初识时动画状态
    int dh_information=13;                      //正常动画开始状态
    int dz_information=-9;                      //打中动画开始状态
    int now_zt=cs_information;                  //实时动画状态

    //动画下一帧的顺序
    public boolean toNext()
    {
        if(now_zt>0)
        {
            //正常动画
            now_zt--;
```

```
        if(now_zt==cs_information)
        {
           //没打中
           return true;
        }
    }
    else
    {
       //打中动画
       if(now_zt<0)
       {
          now_zt++;
       }
    }
    return false;
}
//地鼠出现动画
public void thShow()
{
   now_zt=dh_information;
}
//打中的判定
public boolean hit_pd()
{
   if(now_zt>cs_information)
   {
      now_zt=dz_information;
      return true;
   }
   return false;
}
```

8.4 任务实施（二）

8.4.1 任务描述（Tetris）

俄罗斯方块（Tetris）是由阿列谢·帕基特诺发明的，它在一个 $m×n$ 的矩形框内进行，矩形框的顶部会随机的出现一个由四个小方块组成的砖块，每过一段时间，就会下落一格，直到它碰到底部，然后再过一段时间下落另一个砖块，依次进行，砖块形状是随机出现的。如果底部砖块铺满，则消去铺满的层从而得到相应设置的分数，当砖块到达顶部的时候，游戏结束，整个游戏过程伴有背景音乐。

8.4.2 界面设计

俄罗斯方块游戏界面主要包括游戏容器、下坠砖块、待下坠砖块、右侧是速度和得分。游戏结束后，会弹出另一个界面，可以选择继续或退出游戏。具体界面如图8-2所示。

8.4.3 代码实现

（1）MainActivity.java 文件。

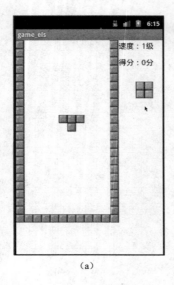 

(a) (b)

图 8-2 俄罗斯方块游戏界面

(a) 游戏主界面；(b) 游戏结束界面

```java
package com.example.game_els;
import android.os.Bundle;
import android.app.Activity;
import android.util.DisplayMetrics;
import android.view.KeyEvent;
import android.view.Menu;
public class MainActivity extends Activity {
    gameview gv;
    protected void onCreate(Bundle savedInstanceState) {
        super.onCreate(savedInstanceState);
        //得到当前窗体的宽高
        DisplayMetrics dm=new DisplayMetrics();
        //用 DM 存储窗体初始化信息
        getWindowManager().getDefaultDisplay().getMetrics(dm);
        //分别取出当前窗体的宽高
        int windows_width=dm.widthPixels;
        int windows_height=dm.heightPixels;
        gv=new gameview(MainActivity.this,windows_width,windows_height);
        setContentView(gv);
    }
    @Override
    public boolean onCreateOptionsMenu(Menu menu) {
        // Inflate the menu; this adds items to the action bar if it is present.
        getMenuInflater().inflate(R.menu.main, menu);
        return true;
    }
    @Override
    public boolean onKeyDown(int keyCode, KeyEvent event) {
        // TODO Auto-generated method stub
```

```
        gv.onKeyDown(keyCode, event);
        return super.onKeyDown(keyCode, event);
    }
}
```

(2) gameview.java 文件。

```
package com.example.game_els;
import javax.crypto.spec.IvParameterSpec;
import android.content.Context;
import android.graphics.Bitmap;
import android.graphics.Canvas;
import android.graphics.Color;
import android.graphics.Matrix;
import android.graphics.Paint;
import android.graphics.drawable.BitmapDrawable;
import android.os.Handler;
import android.view.KeyEvent;
import android.view.MotionEvent;
import android.view.View;
public class gameview extends View {
    /*初始化:1)定义变量;2)初始化程序*/
    /*1)定义变量*/
    //定义方块的样子
    int shapes[][][]={
        //棒形
            {
                {0,0,0,0,1,1,1,1,0,0,0,0,0,0,0,0},
                {0,1,0,0,0,1,0,0,0,1,0,0,0,1,0,0},
                {0,0,0,0,1,1,1,1,0,0,0,0,0,0,0,0},
                {0,1,0,0,0,1,0,0,0,1,0,0,0,1,0,0}
            },
        //田字形
            {
                {0,1,1,0,0,1,1,0,0,0,0,0,0,0,0,0},
                {0,1,1,0,0,1,1,0,0,0,0,0,0,0,0,0},
                {0,1,1,0,0,1,1,0,0,0,0,0,0,0,0,0},
                {0,1,1,0,0,1,1,0,0,0,0,0,0,0,0,0},
            },
        //Z形
            {
                {1,1,0,0,0,1,1,0,0,0,0,0,0,0,0,0},
                {0,1,0,0,1,1,0,0,1,0,0,0,0,0,0,0},
                {1,1,0,0,0,1,1,0,0,0,0,0,0,0,0,0},
                {0,1,0,0,1,1,0,0,1,0,0,0,0,0,0},
            },
        //反Z行
            {
                {0,1,1,0,1,1,0,0,0,0,0,0,0,0,0,0},
                {1,0,0,0,1,1,0,0,0,1,0,0,0,0,0,0},
                {0,1,1,0,1,1,0,0,0,0,0,0,0,0,0,0},
                {1,0,0,0,1,1,0,0,0,1,0,0,0,0,0,0},
```

```java
        },
    //L 形
        {
            {1,0,0,0,1,0,0,0,1,1,0,0,0,0,0,0},
            {1,1,1,0,1,0,0,0,0,0,0,0,0,0,0,0},
            {1,1,0,0,0,1,0,0,1,0,0,0,0,0,0,0},
            {0,0,1,0,1,1,1,0,0,0,0,0,0,0,0,0}
        },
    //反 L 形
        {
            {0,1,0,0,0,1,0,0,1,1,0,0,0,0,0,0},
            {1,0,0,0,1,1,1,0,0,0,0,0,0,0,0,0},
            {1,1,0,0,1,0,0,0,1,0,0,0,0,0,0,0},
            {1,1,1,0,0,0,1,0,0,0,0,0,0,0,0,0}
        },
    //T 形
        {
            {0,1,0,0,1,1,1,0,0,0,0,0,0,0,0,0},
            {1,0,0,0,1,1,0,0,1,0,0,0,0,0,0,0},
            {1,1,1,0,0,1,0,0,0,0,0,0,0,0,0,0},
            {0,1,0,0,1,1,0,0,0,1,0,0,0,0,0,0}
        }
};
    int state;                              //1 表示游戏结束,0 游戏进行中
    int x,y;                                //方块坐标
    int score;                              //得分
    int speed;                              //速度
    int map[][]=new int[13][23];            //画面
    int nowtype;                            //当前方块样子
    int nowstate;                           //当前方块形态
    int nexttype;                           //下一个方块的样子
    int nextstate;                          //下一个方块的形态
    int size;
    playsound ps;
    public gameview(Context context,int width,int height) {
        super(context);
        //构造函数中:2)初始化程序
        ps=new playsound(context);
        score=0;
        speed=1;
        state=0;
        nexttype=(int) (Math.random()*7);
        nextstate=(int) (Math.random()*4);
        size=width/17;
        //产生图
        newmap();
        //产生墙
        newwall();
        //产生方块
        newblock();
        //主动下落
```

```java
            ps.bgmusic();
            h.post(r);
    }
    private Bitmap tpzh(int resid) {
        // 将图片进行等比例缩放
        //先获取资源图片信息
        Bitmap y_bitmap=((BitmapDrawable)(getResources().getDrawable(resid))).getBitmap();
        //获取原始图片的长宽信息
        int y_width=y_bitmap.getWidth();
        int y_height=y_bitmap.getHeight();
        //等比例缩放
        Matrix matrix=new Matrix();
        //设置缩放的比例 s_width  s_height
        float s_width=(float)size/y_width;
        float s_height=(float)size/y_height;
        //设置缩放比例
        matrix.postScale(s_width, s_height);
        //利用比例信息缩放原始图片
        Bitmap x_bitmap=Bitmap.createBitmap(y_bitmap, 0, 0, y_width, y_height, matrix, true);
        return x_bitmap;
    }
    private void newblock() {
        // 产生新的俄罗斯方块 :把下一个状态给当前,再产生新的下一个
        nowtype=nexttype;
        nowstate=nextstate;
        nexttype=(int) (Math.random()*7);
        nextstate=(int) (Math.random()*4);
        x=4;
        y=0;
        if(gameover(x, y))
        {
           ps.sound(4);
           state=1;
           h.removeCallbacks(r);
           invalidate();
        }
    }
    private void newwall() {
        // 将MAP中信息墙位置信息转化成2
        for(int i=0;i<12;i++)
        {
           map[i][21]=2;
        }
        for(int j=0;j<22;j++)
        {
           map[0][j]=2;
           map[11][j]=2;
        }
    }
```

```java
private void newmap() {
    // 最初图,设置全 0
    for(int i=0;i<12;i++)
    {
        for(int j=0;j<22;j++)
        {
            map[i][j]=0;
        }
    }
}
Handler  h=new Handler();
/*俄罗斯方块中的操作 1)主动;2)被动*/
/*1)多线程中实现主动下落*/
Runnable r=new Runnable() {
    @Override
    public void run() {
        // 多线程中主动操作
        if(!blow(x, y+1, nowtype, nowstate))
        {
            editmap(x, y, nowtype, nowstate);

            newblock();
        }
        else
        {
            y=y+1;
        }
        deline();
        postInvalidate();
        h.postDelayed(r, 800-(speed-1)*50);
    }
};
/*2)通过按键实现被动操作*/
@Override
public boolean onKeyDown(int keyCode, KeyEvent event) {
    // 处理键盘的按下操作
    switch(keyCode)
    {
    case KeyEvent.KEYCODE_DPAD_DOWN:down();break;
    case KeyEvent.KEYCODE_DPAD_LEFT:left();break;
    case KeyEvent.KEYCODE_DPAD_RIGHT:right();break;
    case KeyEvent.KEYCODE_DPAD_UP:up();break;
    }
    return super.onKeyDown(keyCode, event);
}
private void up() {
    // 变形操作:先检测碰撞
    int temp=nowstate;
    nowstate=(nowstate+1)%4;
    if(!blow(x, y, nowtype,nowstate ))
    {
        //碰撞了,恢复原状态
```

```java
            nowstate=temp;
        }
        else
        {
            ps.sound(1);
        }
        invalidate();
    }
    private void right() {
        // 方块右移:先检测碰撞
        if(blow(x+1, y, nowtype, nowstate))
        {
            //没碰撞则坐标变化
            x=x+1;
            ps.sound(1);
        }
        invalidate();
    }
    private void left() {
        //方块左移:先检测碰撞
        if(blow(x-1, y, nowtype, nowstate))
        {
            x=x-1;
            ps.sound(1);
        }
        invalidate();
    }
    private void down() {
        // 加速下滑:先检测碰撞
        if(blow(x, y+1, nowtype, nowstate))
        {
            //能下落,则下落
            y=y+1;
            ps.sound(1);
        }
        else
        {
            //不能下落
            //把当前方块变成已落块
            editmap(x, y, nowtype, nowstate);
            //产生新的方块
            newblock();
        }
        invalidate();
    }
    /*碰撞判定*/
    public boolean blow(int a,int b,int ba,int bb)
    {
        //a,b 对应图坐标,ba bb 俄罗斯方块样子,用真表示没碰,假表示碰撞
        for(int i=0;i<4;i++)
        {
            for(int j=0;j<4;j++)
```

```
            {
                //当map中和shapes中都有值了,ba是当前类型,bb是当前存在形态
                //判定是撞墙或者撞方块了
                if((shapes[ba][bb][i*4+j]==1&&map[a+j+1][b+i]==2)
                    ||(shapes[ba][bb][i*4+j]==1&&map[a+j+1][b+i]==1))
                {
                    return false;
                }
            }
        }
        return true;
    }
    /*俄罗斯方块消行问题*/
    public void deline()
    {
        int c=0;//计数,某行1数量为10消行
        int k=0;//确定消几行
        for(int i=0;i<22;i++)
        {
            for(int j=0;j<12;j++)
            {
                //逐行查找1
                if(map[j][i]==1)
                {
                    c=c+1;
                    if(c==10)
                    {
                        //c=10 消除改行;
                        k=k+1;
                        for(int m=i;m>0;m--)
                        {
                            for(int n=0;n<11;n++)
                            {
                                map[n][m]=map[n][m-1];
                            }
                        }
                    }
                }
            }
            c=0;
        }
        if(k>0)
        {
            ps.sound(2);
        }
        switch(k)
        {
            case 1:score=score+100;break;
            case 2:score=score+300;break;
            case 3:score=score+800;break;
            case 4:score=score+1500;break;
        }
```

```
       int t=speed;
       speed=score/10000+1;
       if(t!=speed)
       {
          ps.sound(3);
       }
    }
    /*俄罗斯方块到达后的操作:将方块三维数组中信息 写入到图形中*/
    public void editmap(int a,int b,int ba,int bb)
    {
       ////a,b 对应图坐标,ba bb 俄罗斯方块样子坐标,用真表示没碰,假表示碰撞
       //俄罗斯方块信息压入MAP图中
       int k=0;
       for(int i=0;i<4;i++)
       {
          for(int j=0;j<4;j++)
          {
             if(map[a+j+1][b+i]==0)
             {
                map[a+j+1][b+i]=shapes[nowtype][nowstate][k];
             }
             k++;
          }
       }
    }
    /*游戏结束判定*/
    public boolean gameover(int x,int y)
    {
       if(blow(x, y, nowtype, nowstate))
       {
          return false;
       }
       return true;
    }
    /*看到图像问题*/
    protected void onDraw(Canvas canvas) {
       // 画整个俄罗斯方块图形
       //1)引入要画的图片资源
       //墙使用图片
       Bitmap bm1=tpzh(R.drawable.a2);
       //使用中俄罗斯方块
       Bitmap bm2=tpzh(R.drawable.a6);
       //已落下的方块
       Bitmap bm3=tpzh(R.drawable.a1);
       //下一个方块
       Bitmap bm4=tpzh(R.drawable.a5);
       //成功的图片
       Bitmap bm5=((BitmapDrawable)getResources().getDrawable(R.drawable.win_
panel)).getBitmap();
       //失败的图片
       Bitmap bm6=((BitmapDrawable)getResources().getDrawable(R.drawable.lose_
panel)).getBitmap();
```

```java
            //重玩按钮
            Bitmap  bm7=((BitmapDrawable)getResources().getDrawable(R.drawable.renew)).
getBitmap();
            //退出按钮
            Bitmap  bm8=((BitmapDrawable)getResources().getDrawable(R.drawable.quit)).
getBitmap();
            if(state==0)
            {
                /*画游戏进行中的样子*/

                for(int j=0;j<22;j++)
                {
                    for(int i=0;i<12;i++)
                    {
                        //画墙
                        if(map[i][j]==2)
                        {
                            canvas.drawBitmap(bm1, i*size, j*size, null);
                        }
                        //画已落的方块
                        if(map[i][j]==1)
                        {
                            canvas.drawBitmap(bm3, i*size, j*size, null);
                        }
                    }
                }

                for(int i=0;i<16;i++)
                {
                    if(shapes[nowtype][nowstate][i]==1)
                    {
                        //画使用的方块
                        canvas.drawBitmap(bm2, (i%4+1+x)*size, (i/4+y)*size, null);
                    }
                    if(shapes[nexttype][nextstate][i]==1)
                    {
                        //画下一个方块
                        canvas.drawBitmap(bm4, (i%4+13)*size, (i/4+5)*size, null);
                    }
                }
                //确定得分和速度
                Paint p=new Paint();
                p.setColor(Color.BLUE);
                p.setTextSize(size-3);
                canvas.drawText("速度:"+speed+"级", 12*size, 1*size, p);
                canvas.drawText("得分:"+score+"分", 12*size, 3*size, p);
            }
            else
            {
                Paint p=new Paint();

                if(score>=100000)
```

```
            {//成功
                canvas.drawBitmap(bm5, 100, 200,null);
                p.setColor(Color.RED);
                p.setTextSize(30);
                canvas.drawText("恭喜您！您的得分"+score+"分", 100, 350, p);
                //60*25:200-260 400-425,ontouhchevent
                canvas.drawBitmap(bm7, 150, 400,null);
                canvas.drawBitmap(bm8, 250, 400,null);
            }
            else
            {//失败
                canvas.drawBitmap(bm6, 100, 200,null);
                p.setColor(Color.RED);
                p.setTextSize(30);
                canvas.drawText("很遗憾！您的得分"+score+"分", 100, 350, p);
                canvas.drawBitmap(bm7, 150, 400,null);
                canvas.drawBitmap(bm8, 250, 400,null);
            }
        }
        super.onDraw(canvas);
    }
    @Override
    public boolean onTouchEvent(MotionEvent event) {
        // 判定单击图片按钮
        int mx,my;//接收单击位置坐标
        mx=(int) event.getX();
        my=(int) event.getY();
        if((mx>=150&&mx<=210)&&(my>=400&&my<=425))
        {
            //游戏重新初始化
            //构造函数中:2)初始化程序
            score=0;
            speed=1;
            state=0;
            nexttype=(int) (Math.random()*7);
            nextstate=(int) (Math.random()*4);
            //产生图
            newmap();
            //产生墙
            newwall();
            //产生方块
            newblock();
            //主动下落
            h.post(r);
        }
        if((mx>=250&&mx<=310)&&(my>=400&&my<=425))
        {
            //退出程序:利用android系统来杀死进程
            android.os.Process.killProcess(android.os.Process.myPid());
        }
        return super.onTouchEvent(event);
    }
```

}

（3）playsound.java。

```java
package com.example.game_els;
import java.io.IOException;
import java.util.HashMap;
import android.content.Context;
import android.media.AudioManager;
import android.media.MediaPlayer;
import android.media.SoundPool;
public class playsound {
    //背景音乐:比较长,可以反复播放,mediaplay
    //音效:比较短,快速触发播放:soundpool （一般播放7秒以内的声音）
    Context mc;                                          //上下文 ,activity
    MediaPlayer music=new MediaPlayer();                 //播放背景音乐
    SoundPool sound=new SoundPool(8, AudioManager.STREAM_MUSIC, 0);//播放音效
    //soundpool 一般使用hashmap
    HashMap<Integer, Integer> soundmap=new HashMap<Integer, Integer>();
    public playsound(Context c)
    {
        //接收的界面
        mc=c;
        //初始化游戏音效,放入hashmap
        soundmap.put(1, sound.load(mc, R.raw.atom,1));        //按键的声音
        soundmap.put(2, sound.load(mc, R.raw.clear3,1));      //消行的声音
        soundmap.put(3, sound.load(mc, R.raw.levelup,1));     //升级的声音
        soundmap.put(4, sound.load(mc, R.raw.end,1));         //游戏结束的声音
    }
    //播放背景音乐
    public void bgmusic(){
        music=MediaPlayer.create(mc, R.raw.title);
        try {
            music.prepare();
        } catch (IllegalStateException e) {
            // TODO Auto-generated catch block
            e.printStackTrace();
        } catch (IOException e) {
            // TODO Auto-generated catch block
            e.printStackTrace();
        }
        music.start();
        music.setLooping(true);
    }
    //播放音效
    public void sound(int i)
    {
        sound.play(soundmap.get(i), 1, 1, 1, 0, 1);
    }
}
```

参 考 文 献

[1] [美] Dave Smith,Jeff Friesen 著. 赵凯,陶冶,译. Android 开发范例代码大全. 2 版. 北京:清华大学出版社,2014.
[2] 毋建军,徐振东,等. Android 应用开发案例教程. 北京:清华大学出版社,2013.
[3] 李维勇. Android 任务驱动式教程. 北京:北京航空航天大学出版社,2011.
[4] 楚无咎. Android 编程经典 200 例. 北京:电子工业出版社,2013.
[5] 康艳梅,译. Android 应用开发案例精解. 北京:电子工业出版社,2013.